Governing Nano Foods
Principles-Based Responsive Regulation

Governing Nano Foods

Principles-Based Responsive Regulation

EFFoST Critical Reviews #3

Bernd van der Meulen

Harry Bremmers

Kai Purnhagen

Nidhi Gupta

Hans Bouwmeester

Leon Geyer

AMSTERDAM • BOSTON • HEIDELBERG • LONDON
NEW YORK • OXFORD • PARIS • SAN DIEGO
SAN FRANCISCO • SINGAPORE • SYDNEY • TOKYO

Academic Press is an imprint of Elsevier

Academic Press is an imprint of Elsevier
32 Jamestown Road, London NW1 7BY, UK
The Boulevard, Langford Lane, Kidlington, Oxford, OX5 1GB, UK
Radarweg 29, PO Box 211, 1000 AE Amsterdam, The Netherlands
225 Wyman Street, Waltham, MA 02451, USA
525 B Street, Suite 1900, San Diego, CA 92101-4495, USA

First published 2014

British Library Cataloguing in Publication Data
A catalogue record for this book is available from the British Library

Library of Congress Cataloging-in-Publication Data
A catalog record for this book is available from the Library of Congress

ISBN: 978-0-12-420156-9

For information on all Academic Press publications
visit our website at **store.elsevier.com**

This book has been manufactured using Print On Demand technology. Each copy is produced to
order and is limited to black ink. The online version of this book will show color figures where
appropriate.

CONTENTS

PREFACE AND ACKNOWLEDGEMENTS

This book is an updated and much revised version of a study that has been written to contribute to discussions at the FAO Round Table Sessions at NANOAGRI 2010 in São Pedro, Brazil.[1] UN Food and Agriculture Organization (FAO) requested the initial study. However, the analyses and opinions expressed are those of the authors. They do not necessarily reflect FAO's positions. FAO holds the copyrights of the 2010 study, but authorised the authors to publish the information contained therein. This publication is endorsed by the European Federation of Food Science & Technology (EFFoST) and the Global Harmonization Initiative (GHI).

This book primarily targets at an interdisciplinary audience. This is why the language and content of this book largely deviates from what an audience socialised in the respective science might expect from such a topic. We tried to strike a balance between presenting the core issues of the respective disciplines in a language understandable to an interdisciplinary audience on the one hand, while displaying the main topics from the respective disciplines related to the topic of the book in a most comprehensive way on the other. It will be on the reader to judge if we succeeded.

We are grateful for the trust FAO has placed in the food law team at Wageningen University. The following experts have contributed in various ways to this study:

Gerrit Alink (Wageningen University), Sourav Bhattacharjee (Wageningen University), Arnout Fisher (Wageningen University), Scott Geyer (Massachusetts Institute of Technology), Frans Kampers (Wageningen University), Maryvon Noordam (RIKILT Institute of Food Safety), Onno Omta (Wageningen University) and Merel van der Ploeg (Wageningen University). Other people who helped to make it possible to produce this study are Patrick van Veenendaal and

---

[1]See <www.nanoagri2010.com> and Caue Ribeiro et al., 2010. International Conference on Food and Agriculture Applications of Nanotechnologies, São Pedro.

Ingeborg Oude Lansink at the Liaison Office of Wageningen University.

FAO has initiated development of a publication in collaboration with Monash University on Nanotechnology Applications in Food and Agricultural Sectors: Principles and Guidance for Food Safety Regulation. With the permission of FAO, parts of the text from this draft publication have been included in the appendix of this paper.

Bernd van der Meulen

Wageningen 15 October 2013

EXECUTIVE SUMMARY

The objective of this book is to help to test and improve existing regulatory infrastructures — or if need be to build new ones — in such a way that they can ensure food safety in general in the face of the challenges posed by nanotechnologies.

This book argues for a principles-based regulatory framework for nanotechnologies. In contrast to most conventional regulatory wisdom that identifies gaps in existing regulatory systems with a view of filling them with detailed rules specifically related to the risks emerging from nanotechnology,[1] this book subscribes to research on the ineffectiveness of such rule-based regulation. Instead, and in line with insights from research mainly from the sector of financial regulation,[2] we argue for a more responsive and thus effective principles-based framework to deal with nanotechnology in the food sector. Such a framework may provide an instrument to benchmark existing regimes and to design new or improved structures.

This study builds on existing principles of food safety regulation and the interpretation of these principles at national and regional levels. The study works from the bottom-up to build a regulatory system for addressing the risks connected to nanotechnology applications in food production, despite the fact that profound knowledge on the impact of nanotechnologies is still lacking. The primary objective of the proposed regulatory system is to protect the health and safety of consumers of foods that are likely to contain substances obtained by nanoscale chemosynthesis or engineered nanoparticles. This relates not only to substances included in the food but also to substances that may leave traces in food due to their use at various stages of the food chain. Examples of such substances may be pesticides, equipment and packaging materials.

---

[1]See Appendix C.
[2]See *inter alia* Julia Black, Martyn Hopper and Christa Band, 2007. Making a success of principles-based regulation. Law and Financial Markets Review, 191–206.

There are existing and generally agreed principles on which regulatory approaches to food safety regulation are based. The proposed structure chooses responsibility including liability of food businesses as an underlying principle. Likewise, nanotechnologies applied to foods — nanomaterial in packaging or nanoproducts used in food preparation — may require appropriate safety studies. The core of nanofood safety regulation is a case-by-case premarket approval requirement for foods to which nanotechnologies have been applied. Where premarket approval is required, the sponsor of the product to which nanotechnology has been applied must provide methods of detection and scientific proof of safety of the product at issue. The scientific community and public authorities must develop and agree on methods of risk assessment. To ensure that this principle can be implemented in the context of problems stemming from the foods applications of nanotechnology, registration is desirable of at least all those businesses that bring such foods to the market. Furthermore businesses are required to have systems in place that enable the traceability of such foods. To protect property rights and enable food businesses to benefit from approved applications, applicants should enjoy an exclusive right for a certain period from the moment of approval on. After this exclusivity period, the approval should become generic to avoid obstruction of innovation and unnecessary suffering that may result from a repetition of safety testing on humans or animals.

The leading risk assessment authorities and the international community are called upon to come to understandings regarding cooperation and mutual recognition that ensure world-wide market access for approved applications of nanotechnologies in the food sector and protect businesses from the need to submit to multiple approval procedures. Taken together the proposed measures constitute a regulatory framework capable of ensuring the safety of foods to which nanotechnology has been applied.

ABBREVIATIONS

ADI	acceptable daily intake
BSE	bovine spongiform encephalopathy
CAC	Codex Alimentarius Commission (FAO/WHO)
CCP	critical control point
CFR	Code of Federal Regulations (USA)
EFFL	European Food and Feed Law Review
EFSA	European Food Safety Authority
ENMs	engineered nanomaterials
ENPs	engineered nanoparticles
EPA	Environmental Protection Agency (USA)
EU	European Union
FAO	Food and Agriculture Organization (UN)
FCM	food contact material
FDA	Food and Drug Administration (USA)
FFDCA	Federal Food, Drug and Cosmetic Act (USA)
FSA	Food Standards Agency (UK)
FSAI	Food Safety Authority of Ireland
FSANZ	Food Standards Australia New Zealand
GATT	General Agreement on Tariffs and Trade (WTO)
GI-tract	gastrointestinal tract
GMO	genetically modified organism
GRAS	generally recognized as safe (USA)
HACCP	hazard analysis and critical control points
ICCPR	International Covenant on Civil and Political Rights (UN)
ICESCR	International Covenant on Economic, Social and Cultural Rights (UN)
INS	International Numbering System
ISO	International Organization for Standardization
JECFA	Joint FAO/WHO Expert Committee on Food Additives
JEMRA	Joint FAO/WHO Expert Meetings on Microbiological Risk Assessment
JMPR	Joint FAO/WHO Meeting on Pesticide Residues
NM	nanomaterials
nm	nanometre

NP	nanoparticles
NT	nanotechnology
MRL	maximum residue limit
OECD	Organisation for Economic Co-operation and Development
PARNUTS	foods for particular nutritional uses
PHS Act	Public Health Services Act (USA)
RA	risk assessment
RDA	recommended daily allowance
RFID	radio frequency identification
SPS Agreement	Agreement on the Application of Sanitary and Phytosanitary measures (WTO)
UK	United Kingdom
UN	United Nations
USA	United States of America
USC	United States Code
USDA	Department of Agriculture (USA)
WHO	World Health Organization
WTO	World Trade Organization

CHAPTER 1

Introduction

1.1 WHY REGULATE THE APPLICATION OF NANOTECHNOLOGIES IN FOOD?

Regulatory intervention in product markets such as trade in foods is based on the insight of the imperfection of markets.[1] If nanotechnologies would be subject to the humble forces of markets alone, the risk that negative externalities and information asymmetry would lead to unwanted outcomes would be too high to leave nanotechnology unregulated. However, highly regulated markets may face a detriment in a competition with laxer-regulated markets in the world (so-called Delaware effect[2]). This would mean that businesses that offer nanofoods may be more likely to offer their products in legal regimes where they face a more favourable regulatory framework. The stricter regulation would cut consumers off from getting access to the benefits such a new technology would provide. If the regulated market is big enough, however, strict regulation can also lead to a form of *de-facto* standard setting, which drives other, laxer-regulated regimes to comply with the regulatory standard of the highly regulated regime (California[3]- or Brussels Effect[4]). Hence, regulatory regimes also compete for effectiveness of their methodology across borders. In any way, regulatory regimes shall have an interest in finding an optimal level of regulation, which on the one hand provides for the safety of consumers but gives them access to new technologies, which have the potential to make their life better, on the other. Not regulating nanofoods at all is hence not a viable option, but their prohibition is also not the right way to go. Like in any kind of product safety regulation, the main dilemma for policy makers in regulation of nanofoods is to find the optimal level of regulation in product safety.

---

[1]See Arcuri, 2000.
[2]Coffee, 1987, 761–763.
[3]Vogel, 1995.
[4]Bradford, 2012.

In the food and agricultural sectors, regulatory intervention is framed in human rights obligations, which addresses these shortfalls of imperfect markets from a legal perspective. The vast majority of states in the world has committed themselves to the progressive realisation of the human right to adequate food. As per 10 October 2013, 160 states are party to the United Nations (UN) International Covenant on Economic, Social and Cultural Rights (ICESCR).[5] State parties' responsibility regarding adequate food encompasses diverse dimensions. In order to realise the right to food, states have to solve the dilemma to find an optimal level of regulation: States should on the one hand regard favourably technological and other developments that may contribute to the nutrition of the population in general and of vulnerable groups in particular.[6] On the other, they shall protect[7] people from adverse substances[8] in food.[9]

Nanotechnologies may be relevant from both perspectives: the perspective of providing the world with improved possibilities for sustainable agricultural production to better meet nutritional needs, but also the perspective of hazards and risks from which consumers should be protected. The challenge for governments is to strike the right balance in creating, enabling and protecting regulatory environments.

1.2 APPLICATIONS OF NANOTECHNOLOGIES IN FOOD AND AGRICULTURE

In recent years, properties of substances and processes at nanoscale command scientific attention within a converging field of physics, chemistry, biology, materials sciences and other disciplines. This attention leads to increased understanding regarding existing products and processes, but also to the development of new technologies and

---

[5]See <http://treaties.un.org/Pages/ViewDetails.aspx?src = TREATY&mtdsg_no = IV-3&chapter = 4&lang = en>.
[6]Article 11(2)(a) ICESCR.
[7]On the state obligations to respect, *protect* and fulfil, see General Comment 3 on the nature of State parties' obligations, of the UN Committee on Economic, Social and Cultural Rights.
[8]On the concept of adequacy encompassing, food in a quantity and quality sufficient to satisfy the dietary needs of individuals, free from adverse substances and acceptable within a given culture, see General Comment 12 on the right to adequate food.
[9]On the implementation of the human right to adequate food, see also Voluntary Guidelines, to support the progressive realisation of the right to adequate food in the context of national food security, adopted by the 127th Session of the FAO Council November 2004, Food and Agriculture Organization of the United Nations, Rome, 2005.

substances. While there are is controversy about a legal definition of nanomaterial,[10] nanotechnology can be thought of as any technology which either incorporates or employs nanomaterials or involves processes performed at the nanoscale.[11] Nanotechnologies are generally seen as new and fast emerging fields that involve the manufacture, processing and application of structures, devices and systems by controlling shape and size at the nanometre scale.

Nanotechnologies have the potential to impact many aspects of food and agricultural systems, as nano-enabled technologies can be applied along the entire food production chain from production to consumption (e.g. during cultivation – nanopesticides, food processing, packaging, pathogen detection, etc.).

During a FAO/WHO Expert Meeting on the Application of Nanotechnologies in the Food and Agriculture Sectors: Potential Food Safety Implications (June 2009), the following broad categories of nanotechnology applications in the food and feed sector were identified on the basis of the analysis conducted by Chaudhry et al. (2008):[12]

1. Where nanotechnology processes and materials have been employed to develop food contact materials (FCMs). This category includes nanomaterial-reinforced materials, active FCM designed to have some sort of interaction with the food or environment surrounding the food and coatings providing surfaces with nanomaterials or nanostructures.
2. Where food/feed ingredients have been processed or formulated to form nanostructures. This category includes applications that involve processing food ingredients at nanoscale to form nanostructures or enhance taste, texture and consistency of the foodstuffs.
3. Where nano-sized, nano-encapsulated or engineered nanomaterial (ENM) ingredients have been used in food/feed. This category includes nanoscale ingredients, including additives (such as colourants, flavourings and preservatives) and processing aids (including nano-encapsulated enzymes) that can be produced for a variety of uses.

---

[10]See Bowman et al., 2010; D'Silva, 2011.
[11]Her Majesties Government, UK Nanotechnologies Strategy, Small Technologies, Great Opportunities, 2010, p. 6.
[12]Based on the extensive review by Chaudhry et al. (2008), Chau et al. (2007) and FAO/WHO (2009).

4. Biosensors for monitoring condition of food during storage and transportation. This category includes packaging which include indicators.
5. Other indirect applications of nanotechnologies in the food and feed area, such as the development of nanosized agro-chemicals (including fertilizers, pesticides, etc.), or veterinary medicines.

1.3 THE CHALLENGE TO REGULATE NANOTECHNOLOGY IN FOOD

Nanotechnologies are the latest in a series of scientific advances that, when applied to the food supply, can both deliver benefits for consumers *and* challenge the ability of regulators to evaluate and ensure safety. There are three main factors that underlie the challenges.

1. **A wide diversity of potential food applications of nanotechnology.**
   'Nanotechnology' is not one thing but rather a set of tools for manipulating matter at the molecular and atomic level. At this 'nanoscale' familiar substances (like silica, carbon and clay) take on new properties that have a wide range of potential application and each new application of nanotechnology can raise new, application-specific scientific and regulatory questions.
2. **The same novel properties of nanomaterials that can convey benefits are also likely to raise new safety questions.**
   The consensus of scientific opinion is that *nanoscale material cannot be assumed to be safe* based on what is known about the same substance at conventional, bulk scale. At least until there is a better understanding of the safety issues in general, each food application will need to be assessed carefully to ensure safety.
3. **The scientific tools used today to evaluate the safety of conventional-scale food substances may not be fully applicable to nanoscale substances.**
   For example, the expression of dose levels in animal toxicity studies in terms of mass per unit of body weight may not work for nanoscale particles, for which toxicity could be a function of the ratio of surface area to mass rather than the absolute mass of the substance being tested. In addition, the potential for nanoscale materials to be distributed in the body and to interact at the cellular level differently than conventional-scale materials may require adaptation of existing toxicity test protocols or entirely new study designs.

In short, *to* perform the product-specific safety assessments required for nanoscale materials, food safety regulators may need scientific tools that do not presently exist including methods of detection.

These scientific challenges and uncertainties do not mean that the hands of food safety regulators are tied or that they cannot do their job of protecting food safety while enabling the free movement of foodstuffs. The principles underlying existing practices for evaluating and regulating the safety of substances added intentionally to food provide a strong basis for moving forward. This study highlights the benefits of a food safety framework which is based on such a principles-based regulation instead of rules-based. We will also highlight the limits and possible adjustments of such a framework (i.e. a functional definition, see Appendix B).

1.4 OBJECTIVES, METHODS AND SCOPE OF THE STUDY

This study highlights the benefits of a food safety framework which is principles based. Given the gaps in knowledge and the uncertainties regarding risks of food applications of nanotechnologies, we argue that its regulation shall be based on principles rather than on rules. We build our claim on the insights of research formulated by Julia Black on principles-based regulation, which aims to cope with the limitations of the capacity of traditional rule-based regulation to manage the challenges arising in complex markets with uncertain risks.[13] Principles-based regulation, as developed in financial market regulation, consists of three elements:[14]

1. broad-based standards in preference to detailed rules
2. outcomes-based regulation
3. increasing senior management responsibility.

One main element is the switch in perspective of regulation: Rules-based regulation tries to prescribe the content of the regulated behaviour to a maximum extent, avoiding regulatory gaps that leave room for interpretation. In principles-based regulation, however, principles deliberately allow for such gaps, which are then filled by application: 'Regulatory conversations as to the meaning and application of the rules take centre stage as their meaning and application is elaborated

[13]Black, 2008, 425.
[14]See Black et al., 2007, 191.

on in iterated communications between the regulator and the regulated'.[15] Principles-based regulation hence finds itself in the middle between regulation and de-regulation as it re-frames 'the regulatory relationship from one of directing and controlling to one based on responsibility, mutuality and trust'.[16] It makes use of strategies from 'new governance'[17] or 'decentred regulation',[18] which comprise bargaining theories, behavioural insights and other facets of social life in the regulatory framework. In this sense, it is far from a classical top-down model of regulation.

> The decentred, or polycentric, analysis of regulation has three dimensions: organisational, conceptual and strategic. Organisationally, it draws attention away from individual regulatory bodies, be they at the national or global level, and emphasises instead the multitude of actors which constitute a regulatory regime in a particular domain. Conceptually, the decentring analysis has a particular understanding both of the nature of the regulatory problem and the nature of state-society and intra-state and intra-society relationships. It emphasises the existence and complexity of interactions and interdependencies between social actors, and between social actors and government in the process of regulation. It has a dialectical conception of the regulatory relationships, in which regulator and regulatee are at once autonomous of and dependent on each other. It also rejects the distinction between public and private: both state and non-state actors engage in the function of regulation, both separately and in different types of interrelationship, and indeed state actors may be regulated by non-state actors. The third dimension is strategic or functional. The hallmarks of the regulatory strategies which are engaged in decentred regulation are that they are hybrid: combining governmental and non-governmental actors; multi-faceted: using a number of different strategies simultaneously or sequentially, and indirect: co-ordinating, steering, influencing and balancing interactions between actors and creating new patterns of interaction which enable social actors to organise themselves. Decentred regulation thus engages the strategies of 'smart regulation' or 'new governance' which have been described in a wide range of writings on regulation, though does not see these as the sole preserve of the state.[19]

To make these insights from financial market regulation regime work for the market of nanofoods, however, the theory of principles-based regulation requires adjustment. In financial markets regulation, such a conversation takes place between senior management of banks,

---

[15]Black, 2008, 432.
[16]Black, 2008, 430–434.
[17]Ford, 2008.
[18]Black, 2001.
[19]Black, 2008, 431 (footnotes omitted).

insurance and other players on the financial markets in order to fill these principles with a meaning. In nanofoods, such as in any other area of product safety regulation in the European Union (EU), these conversations take place between scientists, businesses and regulators.[20]

Rules-based regulation of nanotechnologies would, for example, attempt to capture the concept of nanofoods in a clear cut legal definition. Principles-based regulation, by contrast, would rely on a deliberative process between regulators and stakeholders to give meaning to the scope of their joint responsibility for the safety of nanofoods. This meaning can develop over time with the increase of scientific understanding. This last characteristic makes the system responsive.

Rules providing definitions include and exclude regardless of the consequence in practice. Principles providing broad legal concepts by contrast acquire meaning through application. Conscientious application provides instances[21] that taken together grow into meaning. This preposition serves as a starting point for the following analysis.

Principles-based regulation, as refined for the purposes of this study, hence consists of three elements:

1. broad-based standards in preference to detailed rules
2. outcomes-based regulation
3. increasing responsibility of scientific risk assessors and businesses.

The advantages of principles-based over rules-based regulation may be seen as follows:[22]

1. **Effectiveness**
   Detailed rules are often incapable of preventing misconduct, especially in markets which trade with credence goods.
2. **Durability**
   Regulation that a focus on outcomes is more able to adapt to a rapidly changing market environment than one which is based on prescriptive rules.

---

[20]See Arcuri, 2000.

[21]We avoid the word 'precedent' as using this term would burden the principles-based approach with the meaning that have been attached to this term in Common Law thinking.

[22]Julia Black, 2008. Forms and paradoxes of principles-based regulation. Capital Markets Law Journal 3, 432, summarising the approach of the UK's financial services authority. The text was adjusted to meet the criteria of the market for nanofoods.

3. Accessibility

Principles are far more accessible to risk assessors, senior management and smaller businesses in particular than a bewildering mass of detailed requirements.

4. Fostering substantive compliance

A large volume of detailed provisions can divert attention towards adhering to the letter rather than the spirit of rules, making it less likely that the regulator will achieve its regulatory objective.

It is exactly these reasons that make principles-based regulation particularly attractive in the area of nanofoods, which is characterised by highly credence abilities[23] as well as a steadily changing regulatory environment due to the rapidly increasing knowledge gained from research into nanotechnology.

The focus of this study is on *food safety* only, that is to say protection of consumers' life and health. The protection of the health of consumers is taken as a value in itself.[24] Consumers should be exposed to foods only if these foods can reasonably be considered safe.[25] This study does not approach regulation as a means to influence consumer perception[26] in favour or against nanotechnologies.[27]

Only those applications of nanotechnology which result in consumers being exposed to nanoparticles through their gastrointestinal tract are taken into account. These are in particular substances obtained by nanoscale chemosynthesis, products containing engineered nanoparticles (ENPs) and products that have been exposed to such substances and particles (together also referred to as nanomaterials) in such a way that they may contain residues. Consumers are typically exposed to

---

[23]See on the credence ability of food Bremmers et al., 2013.

[24]In the EU, for example, and in particular in the case law of the ECJ the protection of the health of consumers is taken as a higher value, which generally trumps economic consideration of free trade. See in this respect ECJ, 17 July 1997, Case C-183/95, *Affish BV* v. *Rijksdienst voor de Keuring van Vee an Vlees*, [1997] ECR I-4362, para 43; approved ECJ, 19 April 2012, Case C-221/10P, *Artedogan GmbH* v. *European Commission*, [2012] ECR I-0000 (nyr), para 99.

[25]See for example, Kallet and Klink, 1933.

[26]In its White Paper on Food Safety (COM(1999) 719), the European Commission has taken another position. The Commission sees regaining of consumer trust (in authorities, businesses, science and the food supply) as one of the objectives of the reform of EU food law initiated by the White Paper.

[27]Indeed we believe that governments should not go beyond enabling consumers to make informed choices. We see engineering of opinions even for the best of purposes such as public health as a threat to civil liberties.

such materials when they entered the food in agriculture (nano-engineered pesticides, veterinary drugs or feed) in processing (packaging, utensils and other FCMs) or were added as ingredients. In this study, all products within this scope are loosely referred to as 'nanofoods'. This expression 'nanofoods' must not be seen as indicating a monolithic category, but rather as indicating the outer boundary of a diverse group of products and substances that deserve attention on a case-by-case basis.

From the objective and scope of this book, it follows that other important nanotechnology-related issues such as occupational health and safety (e.g. the safety of researchers and workers), environmental health and safety, fair trade and intellectual property are outside the scope of this study. Also outside the scope of this study is nanoscale mechanical manufacturing, such as nanobots and the use of radio frequency identification (RFID) tracking chips in the food chain and related privacy concerns.

According to the reasons stated above, this book will select various principles of food safety regulation and test them against the possibility to make them work for regulation of nanotechnology. The starting point of analysis is hence the wording of the respective legislative or judicial provisions in international sources. To this end, on the basis of international comparison, we will present an overview of core issues addressed in food safety legislation. However, as this book is based on a contextual understanding, such analysis is not limited to the syntax of positive law. Law is not only geared towards the realisation of individual rights, but also 'to achieve specific goals in concrete situations'.[28] 'The object of scientific research is not so much the legal terminological conceptions, but rather the problems in life they aim to solve'.[29] We hence understand our object of research the 'regulatory framework' to include the hierarchy[30] of hard and soft law: of principles, acts, regulations, guidelines, codes of conduct, implementing

---

[28]Teubner, 1987, 15.

[29]Zweigert and Kötz, 1996, translation by Kai Purnhagen.

[30]In the EU, Regulation (EC) 178/2002 holds a definition of the concept of food law that also expresses the complexity of the legal structure. Article 3(1) of this regulation reads: 'food law' means the laws, regulations and administrative provisions governing food in general, and food safety in particular, whether at community or national level; it covers any stage of production, processing and distribution of food, and also of feed produced for, or fed to, food-producing animals.

policies, scientific policies, conformity assessment requirements, powers of inspection, sanitation and sanctioning and any other tools (by whatever name) that are in place to deal with food safety issues.[31] The framework encompasses provisions of a general nature empowering public authorities and addressing food businesses and moves down to include administrative decisions regarding the marketability of categories of foods and the compliant status of specific foods on the market. We will test whether and to what extend this regulatory framework is able to cope with the specific regulatory risk which evolve from the regulation of nanofoods.

To the extent conventional approaches prove suitable for application in the context of nanotechnologies, contours may emerge of a 'future–proof' structure of food safety law. That is to say, conventional approaches to food law that prove capable to deal with developments unknown at the time they were designed may be expected to be sufficiently flexible or adaptive to deal with other innovations as well.

1.5 STRUCTURE OF THE STUDY

The thinking behind the structure of this study is set out in Appendix A. This appendix presents a structure of food safety regulation summarised in Figure A.4 depicting a model food regulatory system and providing the frame of analysis of this study.

Chapter 2 addresses the underlying generally accepted principle of food safety and the derived requirements from international (World Trade Organization (WTO)) law that national regulatory structures for food nanotechnology should meet. In particular the system should be based on risk analysis including precaution.

The core of this study is Chapter 3 in which the frame of analysis is applied to foods to which nanotechnology has been applied. Chapter 4 singles out a topic that may be of special interest for the regulation of such foods: the application of premarket approval requirements. Chapter 5 provides concepts for legislation specifically addressing foods to which nanotechnology has been applied. In so far – and only in so far – as the safety of such foods cannot be ensured by general food

---

[31]See also, the joint FAO/WHO publication (2003). Assuring Food Safety and Quality. Guidelines for Strengthening National Food Control Systems; Vapnek and Spreij, 2005 (hereafter also referred to as FAO model food law).

safety law, nano-specific requirements are needed. These in turn need clear concepts to delineate their target and scope. In this sense, Chapter 5 provides an excursion into rules-based regulation. Chapter 6 discusses the consequences of the proposed system for businesses in terms of regulatory burdens. It suggests measures to mitigate such burdens. This chapter also searches for the appropriate level of nano-specific regulation. It seems likely that in the short term, measures will be taken at a national level in several countries. In the long run mutual recognition and international regulation may be preferable. Chapter 7 takes the stock of the findings in terms of a responsive system. Chapter 8 concludes the argument of this study by pointing to possible approaches to regulating the application of nanotechnology in the agro-food sector and making recommendations. Two topics addressed in the study — the concept of nanotechnology and existing evaluations of regulatory systems — are further elaborated in the Appendixes B and C. Appendix D quotes from bills addressing foods to which nanotechnology has been applied. Appendix E introduces the research team. For the convenience of the reader, the main text of this study is focussed on the core issues. Discussion of underlying concepts, backgrounds, illustrations and sources is — depending on their length — placed in footnotes and appendixes. The reference list provides hyperlinks to documents available through the Internet.

CHAPTER *2*

Requirements for Food Safety Regulation

2.1 INTRODUCTION

This chapter describes how the three elements of principles-based regulation work together in food safety governance. Section 2.2 introduces the first element consisting of the setting of broad-based standards in food safety governance. Section 2.3 elaborates on two elements of principles-based regulation. It shows how risk analysis puts flesh on the bones of these principles, thereby determining the regulatory outcome of these standards. Section 2.4 elaborates how the need to require scientific data primarily from private risk assessors increases responsibility of private businesses.

2.2 FUNDAMENTAL PRINCIPLES OF FOOD SAFETY GOVERNANCE

There are two main principles in food safety governance, which qualify for general application: the principle of safety (Section 2.2.1) and the principle of science-based decision making (Section 2.2.2).

2.2.1 The Principle of Safety

The main principle of trade in food markets is the prohibition of marketing of unsafe foods. In the EU for example, to this end, Article 14 (1) of Regulation (EC) 178/2002 laying down the general principles and requirements of Food Law (hereinafter General Food Law) stipulates that:

Food shall not be placed on the market if it is unsafe.

Other legislation such as the Canadian Food and Drugs Act[1] or the US-American Federal Food Drug and Cosmetic Act[2] do not carry such

---

[1]C.R.C., c. 870.
[2]FFDCA [21 USC].

an explicit overall command, but contain a catalogue of regulatory measures designed to ensure the safety of foods. Article 2 of the SPS Agreement[3] obliges Members of the WTO to 'ensure that any sanitary or phytosanitary measure is applied only to the extent necessary to protect human, animal or plant life or health, (...)'. From the Codex General Standard on Food Additives can be deduced a general principle of no harm. Only those substances may be added to food for which it has been demonstrated that they can be ingested daily over a lifetime without an appreciable health risk; that is to say that have been assigned an acceptable daily intake (ADI).[4]

Ensuring safety of food is a joint responsibility of the private sector businesses that produce and market food and food ingredients and the government bodies that set and enforce standards to ensure food safety. Under most national systems in the developed world, private businesses have a general duty not to market unsafe foods and food components, with governments being empowered to act to remove unsafe foods from commerce and to take action against private businesses that do not conform to required food safety practice. Ultimately it is up to the regulators to establish an appropriate level of protection.

2.2.2 The Principle of Science-Based Decision Making
Whenever one has to determine the safety of foods, it has to be based on a scientific assessment. In the EU Article 1(2) General Food Law stipulates that a 'strong science base' needs to underpin any 'decision making in matters of food and feed safety'. Likewise, Article 2(2) of the SPS Agreement requires 'that any sanitary or phytosanitary measure' of WTO Members 'is based on scientific principles and is not maintained without sufficient scientific evidence (...)'.

Risk analysis is the generally accepted methodology through which science contributes to the regulation of food safety. It encompasses three steps: risk assessment, risk communication and risk management. The Codex Alimentarius and an increasing number of countries such as the USA, the EU, Australia/New Zealand, Canada and Japan draw from the responsibility of private businesses for food safety the conclusion that sponsors of new substances added to food must provide the

---

[3]Agreement on the Application of Sanitary and Phytosanitary measures (WTO).
[4]Articles 1(1) and 2(b) Codex General Standard for Food Additives, CODEX STAN 192-1995 (lastly revised 2009).

scientific data needed for risk assessment as a basis for risk management decisions regarding market access of the substance at issue.

2.3 RISK ANALYSIS AS A METHOD TO DETERMINE THE REGULATORY OUTCOME

Broad principles require substantiation in order to make them work
for everyday practice in regulatory regimes. According to the principle
of science-based decision making, in food law such a substantiation
has to be based on a scientific risk analysis.

The concept of risk analysis has been elaborated in the *Codex
Alimentarius Procedural Manual*[5] and was taken up by a number of
laws around the world. Risk analysis should follow a structured
approach comprising three distinct but closely linked components: risk
assessment (Section 2.3.1), risk management (Section 2.3.2) and risk
communication (Section 2.3.3).[6] Depending on the context in which
risk analysis is applied, two different types of question are asked from
science:[7]

1. to identify risks and/or
2. to exclude risks.

As we will see below, the type of question at issue has an impact on the
burden of proof and on the consequences of inconclusive risk assessment.

The WTO SPS Agreement seems to regard risk analysis as the
responsibility of the authority that wishes to create a trade barrier.
In such a situation science is asked to positively identify a risk that is
then taken care of by the protective measure. Under premarket
approval schemes introduced here below, the question to science is not
to *identify* risks but to *exclude* them. Only if proof of safety is provided, will the product at issue be allowed to the market.[8]

2.3.1 Risk Assessment

The *Codex Procedural Manual* defines risk assessment as a scientifically based process consisting of four steps: (i) hazard identification,

---

[5]Codex Alimentarius Commission. Procedural Manual, 21st edition, Rome 2013, 106 v.v.
[6]Working Principles for Risk Analysis for Application in the Framework of the Codex Alimentarius, Procedural Manual, 21st edition, pp. 107–113.
[7]On this topic see van der Meulen, 2009a.
[8]On the SPS conformity of such reversal of the burden of proof, see Section 6.2.

(ii) hazard characterisation, (iii) exposure assessment and (iv) risk characterisation.[9]

In international regulatory frameworks, three joint FAO and WHO committees undertake risk assessment: the joint FAO/WHO Expert Committee on Food Additives (JECFA); the Joint FAO/WHO Meeting on Pesticide Residues (JMPR) and the Joint FAO/WHO Expert Meetings on Microbiological Risk Assessment (JEMRA). They advise on maximum (residue) limits for food additives, pesticides and pathogens and on other scientific issues regarding food safety. These three risk assessors are independent from the risk managers. Furthermore, Codex Principles for risk analysis for application by governments calls for the separation of risk assessment and risk management functions to the extent practicable. In the EU, the European Food Safety Authority (EFSA) is charged with risk assessment. Risk management is the responsibility of political bodies such as the European Commission (EC), the Council, the European Parliament and the Member States. EFSA is independent from these authorities but itself holds no powers of risk management. In the USA by contrast, both scientific risk assessment and (administrative and regulatory) risk management are the responsibility of the Food and Drug Administration (FDA) or — for certain types of foods of animal origin — the Department of Agriculture (USDA). In other countries varying degrees of separation of risk assessment and risk management functions are observed. Apparently, the necessity of independence of risk assessment is open to debate.

This model for risk assessment is considered by most experts as adequate to assess the risks related to foods produced using nanotechnologies. However, it is also widely recognised that many gaps exist in the actual knowledge and understanding of the potential hazards deriving from nanotechnologies applied in the agri-food sector and that there is a need for additional tools to support risk assessment of nanomaterials.[10] In particular, the association between toxic effects and dose, as well as the characterisation of nanoparticles, are probably the two most urgent issues awaiting resolution for a better understanding of risk potentially deriving from ENMs. At present, a case-by-case approach to risk assessment of ENMs is being taken. However, risk

[9]Definitions of Risk Analysis Terms Related to Food Safety, Codex Alimentarius Commission, Procedural Manual, 21st edition, Rome 2013, p. 107.
[10]In regulatory regimes where the precautionary principle is recognised, such lack of scientific evidence may justify regulatory intervention.

assessment authorities must prioritise research to generate the required data to fill existing knowledge gaps and develop comprehensive guidelines for risk assessment of ENMs.

One of the main recommendations of the 2009 FAO/WHO Expert Meeting on the Application of Nanotechnologies[11] was to consider the use of a tiered risk assessment approach for application of nano-technologies to food and feed. It should consist of an initial screening level, to characterise the material and to estimate toxicity and exposure or dose−response relationships. It should subsequently progress through more refined and data-intensive tiers, if appropriate. Implementation of such an approach will result in increased knowledge of the relationships between physicochemical characteristics and biological interactions. Ultimately this may enable the prioritisation of types or classes of materials where additional data are likely to be necessary to reduce uncertainties in the risk assessment.

2.3.2 Lacking Data, Inconclusive Risk Assessment and Precautionary Principle

The concept of risk analysis presupposes two things: (i) the existence of scientific data about the potential hazards of a product of sufficient quantity and quality to conduct risk assessment and (ii) conclusive outcomes of the risk assessment. However, especially regulation of new products such as nanomaterials are often characterised through absence of data and uncertainty about potential hazards deriving from it before or even after risk assessment. In such situations, there are several options for regulators:

1. Prohibit the marketing of the products until substantial scientific data is available as it has not yet proven to be safe.
2. Allow the marketing of the product as its unsafeness has not been proven.
3. Allow only conditional marketing of the product (e.g. labelling 'safety has not yet been proven').

Regardless of which of the approaches a regulator shall choose, there are several questions that need answering along the way. If more data is needed for marketing, who has the obligation to acquire it? If marketing is only conditional, what are the conditions? What level of protection is required on such a market? How to deal with conflicting data?

---

[11]FAO/WHO, 2009.

Confusingly, in literature these issues are not always distinguished, but lumped together under the heading 'precaution'. The cause for the confusion is that the expression is used to indicate at least three different notions. Precaution is used (i) to attribute responsibility for providing scientific data needed for risk assessment, (ii) to choose the risk management option after risk assessment has shown the existence of a risk and – last but not least – (iii) as a basis for risk management in cases where risk assessment is inconclusive.

1. **Acquisition of data**

   In some situations the burden to carry out scientific research to acquire the data needed for risk assessment is shifted from public authorities to private parties.[12] In particular in premarket approval schemes, businesses that wish to bring products to the market that are considered (potentially) hazardous are required to provide scientific substantiation of the safety of these products before market access can be granted.

2. **Level of care (protection)**

   Public authorities regulating food safety need to choose a level of care or protection. In case risk assessment identifies a certain food safety risk, a position must be taken whether this risk is acceptable or to what extent it should be reduced or eliminated. Very low limits and zero tolerances set for certain contaminants are sometimes labelled 'precautionary'.[13]

3. **Scientific uncertainty**

   How do risk managers have to proceed if the outcome of risk assessment is inconclusive?

   The Codex Alimentarius explicitly chooses as its foundation that codex standards are based on scientific risk analysis, in particular with regard to their health and safety aspects. Precaution is recognised as an inherent element of risk analysis. However, in situations of scientific uncertainty, as a rule, no Codex standards should be adopted.[14]

   Working Principles For Risk Analysis For Application In The Framework Of The Codex Alimentarius

   When there is evidence that a risk to human health exists but scientific data are insufficient or incomplete, the Codex Alimentarius Commission

---

[12]See for example, EC, Communication on the precautionary principle, COM(2000) 1. More in detail, see van der Meulen et al., 2012.

[13]Otsuki et al., 2001.

[14]The following quote is taken from documents contained in Codex Alimentarius Commission, Procedural Manual, 21st edition, Rome 2013, p. 108.

should not proceed to elaborate a standard but should consider elaborating a related text, such as a code of practice, provided that such a text would be supported by the available scientific evidence.

Precaution is an inherent element of risk analysis. Many sources of uncertainty exist in the process of risk assessment and risk management of food related hazards to human health. The degree of uncertainty and variability in the available scientific information should be explicitly considered in the risk analysis. Where there is sufficient scientific evidence to allow Codex to proceed to elaborate a standard or related text, the assumptions used for the risk assessment and the risk management options selected should reflect the degree of uncertainty and the characteristics of the hazard.

Here we find a significant difference between the Codex Alimentarius and an actual regulatory system. The Codex is a *model* for national legislation. It is not in itself a set of rules to which stakeholders must comply. Where international consensus does not exist, national authorities may still need to take action as appropriate to the national context. The Codex presents a model on the topics where consensus has been reached and leaves other issues open for future consideration.[15] At the national level, the food regulatory framework has to

---

[15]In the same line the EC, after a regulatory review in 2008, took the position that there is no immediate need for new legislation on nanotechnology, and that adequate responses can be developed – especially with regard to risk assessment – by adopting measures, guidelines, etc. under existing legislation (see Understanding Public Debate on Nanotechnologies Options for Framing Public Policy edited by Schomberg and Davies (2010). While, in the absence of a clear consensus on definitions, the preparation of new nano-specific measures is thought to be difficult, and although there continues to be significant scientific uncertainty on the nature of the risks involved, good governance will have to go beyond policy making focused on legislative action. The power of governments is limited by their dependence on the insights and cooperation of societal actors when it comes to the governance of new technologies: the development of a code of conduct, then, is one of their few options for intervening in a timely and responsible manner. The Commission states in the second implementation report on the action plan for Nanotechnologies that 'its effective implementation requires an efficient structure and coordination, and regular consultation with the Member States and all stakeholders' (European Commission, 2009). In this sense, legislators are dependent on scientists' proactive involvement in communicating possible risks of nanomaterials and must steer clear of any legislative actions which might restrict scientific communication and reporting on risk. The ideal is a situation in which all the actors involved communicate and collaborate. Through *codes of conduct*, governments can allocate tasks and roles to all actors involved in technological development, thereby organising collective responsibility for the field. As it is stated in the Consultation Paper on Nanotechnologies published by the European Commission: 'A European Code of Conduct for Responsible Nanosciences and Nanotechnologies Research is part of the European Commission's ambition to promote a balanced diffusion of information on nanosciences and nanotechnologies and to foster an open dialogue, involving the broadest possible range of interested parties'. The starting point of the principles to be set out in the Code of Conduct would be the legal guarantees set out in the Charter on Fundamental Rights, as well as the general principles resulting from relevant international treaties such as the Convention on Human Rights (1950), the Convention on Human Rights and Biomedicine (1997) and the Aarhus Convention for Environment (1998).

encompass the entire area without leaving any issue unregulated. The WTO SPS Agreement acknowledges this dilemma. In Article 5(7) the SPS Agreement stipulates:

> *In cases where relevant scientific evidence is insufficient, a Member may provisionally adopt sanitary or phytosanitary measures on the basis of available pertinent information, including that from the relevant international organisations as well as from sanitary or phytosanitary measures applied by other Members. In such circumstances, Members shall seek to obtain the additional information necessary for a more objective assessment of risk and review the sanitary or phytosanitary measure accordingly within a reasonable period of time.*

Often this is meant by precautionary principle: the justification for risk managers to work from a worst case scenario when scientific risk assessment indicates that health risks may exist but causality relations remain uncertain. Obviously, this approach to precaution is particularly applied in a context where restrictive measures (i.e. barriers to trade) need scientific justification. In cases where proof of safety is required (market access requirements) it will play a less important role or none at all. As long as risk assessment is inconclusive, the required proof of safety has not been provided and access will not be granted.

The scientific situation we are currently facing in some applications of nanotechnology seems to be one of inconclusive risk assessment. We know that food safety hazards are probable but for most food applications of nanotechnology they have not actually been established. This brings us within the ambit of precaution. What we know about the differences between nanomaterials and conventional materials calls for caution, but does not provide much in terms of more detailed risk assessment. This again calls for the development of methodologies of risk assessment to be applied on a case-by-case basis. It will be the harvest of such individual case-by-case analysis rather than grand general rule-based regulation which will provide further guidance for regulators.

2.3.3 Risk Management

Risk management describes the political process that deals with the data acquired from risk analysis. It concerns if and what kind of regulatory decisions shall be taken according to the data available. 'Sanitary measures' as defined in the SPS Agreement are basically risk management measures. According to Annex A(1) of the SPS Agreement:

'[s]anitary or phytosanitary measures include all relevant laws, decrees, regulations, requirements and procedures including, *inter alia*, end product criteria; processes and production methods; testing, inspection, certification and approval procedures; quarantine treatments including relevant requirements associated with the transport of animals or plants, or with the materials necessary for their survival during transport; provisions on relevant statistical methods, sampling procedures and methods of risk assessment; and packaging and labelling requirements directly related to food safety'. In other words sanitary measures encompass both the setting of the regulatory system through general rules (regulation) and the application of the system through individual decisions.

2.3.4 General and Individual Decisions
In order to be able to meet the challenge to ensure food safety, the regulatory system has to address the *business* handling the food, the *food* itself (the product), the *way* in which the businesses deal with the food in production and trade and the *communication* by businesses to consumers about the food. For this underlying structure, see Appendix A.

The regulatory framework not only consists of general rules and principles of an abstract nature but also of decisions applying the rules to specific situations, by granting or withholding approvals, seizure or banning of products and so forth. Such decisions complement the requirements on businesses and make them operational.

2.4 INCREASING RESPONSIBILITY OF BUSINESSES (PRIVATE) RISK ASSESSORS

Involvement of (private) risk assessors into risk management translates into increasing responsibilities of these. Regulatory systems hence need to find ways to incentivise risk assessors and businesses to comply with their responsibilities. As will be laid out in more detail in Section 3.1, one way to ensure such compliance is the introduction of liability. In addition to, and most of the times intertwined with civil liability, possibilities of public enforcement through inspection and – where necessary – sanctions form other incentives. However, regulatory intervention and liability require a benchmark, a certain line when legal consequences are attached to misbehaviour. In legal terms, this is required by the level of care one is expected to fulfil. According to the principle of science-based decision making, such a duty of care has to be based on measurable scientific evidence. The crucial part in particular in the area of nanofoods is that there is no such data available. In the absence of such data, the duty of care will have to be mainly reduced to process documentation, an obligation to keep books and report to the competent authorities as soon as a food risk. Such an obligation will be proposed in Section 3.5.2.

In so far as enforcement is of a punitive nature, accused persons (especially businesses) should enjoy the rights of defence enshrined in the International Covenant on Civil and Political Rights, Article 14 in particular, including the right to be held innocent and the privilege not to testify against himself. For this reason punitive law enforcement may be time consuming. In the face of a food safety incident, the risk of loss of time may not be acceptable. Thus in addition to powers of law enforcement, powers of incident management are needed that address the food safety problem directly and not the behaviour of the alleged perpetrator; at least not in punitive terms. In the exercise of such powers, full cooperation of the responsible persons and businesses can be required.[16] Obviously, the fruits of such cooperation should not be used against the business in a parallel or subsequent punitive procedure as this would seriously jeopardise businesses' willingness to cooperate.[17]

2.5 CONSEQUENCES FOR REGULATION

From a regulatory perspective, the most important challenge the application of nanotechnologies in the agro-food sector poses is the lack of scientific data about potential risks for consumers. We know what we do *not* know rather than what we *do* know. In order to cope with the regulation of nanotechnology successfully, the regulatory regime has to

---

[16]On this topic see FAO/WHO publication (2003). Assuring Food Safety and Quality: Guidelines for Strengthening National Food Control Systems.

[17]The German application of this thought − sec. 44(6) of the Lebensmittel-, Bedarfsgegenstände- und Futtermittelgesetzbuch (the Code on food, food contact materials and feed) − explicitly states that the information provided by the food business operator may not be used against him in criminal proceedings. This provision will undoubtedly stimulate operators to come forward with problems they discover within their organisation. However, there is also the risk that such a provision will be misused to escape punishment for intentional neglect.

be responsive in a sense that it adapts itself to the increasing under-standing of the risks involved that will emerge over time. For the regulatory regime to be responsive, it needs to be open to benefiting from communication processes between regulators and regulated in risk assessment. Principles-based regulation provides a vital form of this kind of regulatory framework, as it provides incentives to put just as much flesh on the bones of a policy goal as there is knowledge available. It invites rather than discourages the free exchange of opinions in procedures of review and litigation. Public authorities − be they administrative or judiciary − must take responsibility to set meaningful examples and precedents by providing ample motivation for each application and interpretation of the system. Developing insights should be shared. To this end, authorities can publish guidance documents and the private sector best practices.

CHAPTER 3

Safety Frame Applied to Food Applications of Nanotechnology

3.1 INTRODUCTION

The requirement of risk analysis discussed in Chapter 2 regards the procedure that must be followed in setting up a food safety regulatory system. It does not say much about the content of such system. Departing from our frame of analysis for food regulatory systems,[1] food safety regulation can be grouped along the lines of the following elements: the food *business*, the food *product* as such, the *process* of production and trade and the *communication* to consumers. In this chapter we will analyse for each of these core elements how they can be applied to foods to which nanotechnology has been applied: the business in Section 3.2, the product in Section 3.3, the process in Section 3.4 and communication in Section 3.5. In Section 3.6 we look at the related topic of FCMs. This chapter concludes in Section 3.7 with a recommendation regarding the regulation of foods to which nanotechnology has been applied.

3.2 FOOD BUSINESS

3.2.1 Responsibility and Liability

For the food safety framework to apply successfully in general and to nanotechnologies applied to foods in particular, it is important to have a solid structure in place holding businesses responsible and liable (under civil or criminal law) for the safety of the foods they bring to market and for compliance with more specific requirements.[2] Some systems rely more on the deterrent effect of liability of businesses for unsafe food,[3] other systems put more emphasis on categorisation and regulation of certain hazards.[4] Obviously a system can rely on liability

---

[1]Set out in Appendix A.
[2]See also Chaudhry et al., 2008.
[3]Such systems seem to fit best in the legal culture of common law countries.
[4]Such systems seem to fit best in the legal culture of civil law countries.

only if it is not too easy to escape through corporate structures or location outside the country at issue.

A basis in responsibility and liability of businesses that bring food to the market is vital for any regulatory system.[5] It ensures that businesses hold a full stake with regard to their products and the consequences these products may cause. This stake may provide an incentive for an active contribution in increasing knowledge and management of safety and in compliance with safety requirements.

3.2.2 Registration
To ensure compliance through inspection and to enforce liability it may be important to know which businesses apply nanotechnologies or import products from nanotechnologies. Registration or licensing schemes may be helpful.

3.3 FOOD PRODUCTS

The frame of analysis set out in Appendix A, identifies six different approaches to food.

A. *The product in concreto*
1. Any food placed on the market must be fit for human consumption in terms of its safety.
B. *Categorical approaches to foods and food ingredients*
2. Certain categories of foods are considered safe and can freely be used;
3. Certain categories of foods are considered unsafe and banned;

---

[5]See for example Article 29 of the FAO model food law:
1. Any person who sells any food that:
   a. has in or upon it any poisonous or harmful substance;
   b. is not wholesome or is otherwise unfit for human consumption;
   c. is adulterated; or
   d. is injurious to human health
      shall be guilty of an offence.
2. In determining whether an article of food is injurious to human health, due regard shall be given not only to the probable effect of such food on the health of a person consuming it, but also to the probable cumulative effect of articles of substantially similar composition on the health of a person consuming such articles in ordinary quantities.
   Comparable notions can be found in the EU in Article 14 and 17 of Regulation (EC) 178/2002; in the USA in Sections 301 and 402 of the Federal Food, Drug and Cosmetic Act [21 USC 331 and 342].

4. Certain categories of foods are presumed unsafe and therefore banned unless/until food businesses prove otherwise.
C. *Approaches to substances in foods*
5. Public authorities set limits to the presence of substances or organisms they consider hazardous.
6. Public authorities apply zero tolerance to the presence of substances or organisms they consider extremely hazardous.

How can these approaches play a role in the regulation of foods to which nanotechnology has been applied?

3.3.1 Food Safety
3.3.1.1 Duty of Care (1)

It seems vital that food businesses are not only responsible for compliance with more specific requirements, but more in general are obliged to take care that the food they bring to the market is safe. An approach that uses the safety of the food as a criterion can be applied to foods to which nanotechnologies have been applied without the need for a definition of nanofood. The concept of safe food is a responsive concept in that each further understanding of the effects of nanotechnology in food gives meaning to the concept of safety without the need to change the concept, in order to bring it into line with new understanding. It is sufficient for the concept to adapt that the state of the art in safety assessment of nanotechnology is made visible to all stakeholders in well-reasoned decisions, rulings or guidance documents.

The Codex Alimentarius defines the concept of food safety[6] as: assurance that food will not cause harm to the consumer when it is prepared and/or eaten according to its intended use.

To be able to comply with safety requirements, businesses applying nanotechnologies, need risk assessment tools to be able to make their judgements. To be able to enforce food safety requirements with regard to foods to which nanotechnologies have been applied, authorities need detection and risk assessment tools. It is the responsibility of science and of sponsors of foods to which nanotechnologies have been applied in cooperation with public authorities to develop and provide methods of detection and risk assessment.

[6]Recommended International Code of Practice General Principles of Food Hygiene, CAC/RCP 1-1969, rev. 4 (2003), p. 7.

3.3.2 Conventional Foods

3.3.2.1 Presumption of Safety (2)

Most systems seem to recognise a wealth of conventional foods[7] considered safe on the basis of them having a history of safe use. Many foods naturally containing nanoparticles (such as milk) will be in this category of foods that can freely be brought to the market. Usually this is not in itself a defined category, but comprises the foods that the regulatory system at issue does not place in another category. Therefore, its relevance to nanotechnologies will mainly depend on the choices made to specifically regulate this topic, to catch it within the scope of other categories or to leave it unregulated. In many systems, non-regulated foods will by default fall within this unsuspected class.

3.3.3 Banned Foods

3.3.3.1 Moratorium (3)

Some products are known or feared to be so hazardous that they are categorically excluded from the food chain. In some parts of the world this is applied for example to BSE[8] risk materials. Some advocate such a moratorium on nanofoods.[9] As will become apparent here below we do not advocate a generic approach to nanofoods, but instead a safety assessment on a case-by-case basis resulting in individual decisions regarding market access of foods to which nanotechnologies have been applied.

3.3.4 Market Access Requirements

3.3.4.1 Approval (4)

In the 1950s, the USA introduced a requirement of premarket approval for food additives. Similar approaches now exist in other parts of the world as well for food additives, food supplements, genetically modified foods, novel foods and some other categories. These categories are considered *a priori* hazardous. Foods falling within the category are banned unless/until for specific foods to the satisfaction of the competent authorities proof of safety has been provided. If a separate category were to be created for foods to which nanotechnologies have been

---

[7]In the USA such foods are called 'non-regulated'.

[8]Bovine Spongiform Encephalopathy (BSE) is a transmissible, neurodegenerative, fatal brain disease of cattle. The disease has a long incubation period of 4–5 years, but ultimately is fatal for cattle within weeks to months of its onset. World Health Organization Fact Sheet.<http://www.who.int/zoonoses/diseases/bse/en/>.

[9]ETC Group. Action Group on Erosion, Technology and Concentration, 2006. Nanotech product recall underscores need for nanotech moratorium: is the magic gone?; ETC Group, The Big Down: Atomtech – Technologies Converging at the Nano-scale, 2003.

applied, this would require a clear definition of the foods falling within this category. Alternatively, such foods may fall within existing categories that require approval (such as food additive or novel food). In the latter situation no specific definition of nanofood would be needed, but instruments to assess whether a given food to which nanotechnology has been applied falls within the given categories. To apply approval procedures to foods to which nanotechnology has been applied, it is necessary that scientific risk assessment of these foods is possible and that the regulator agrees to accept certain methods and their outcomes.

As elaborated below, to do justice to the basic principle of food safety, we advocate application of premarket approval requirements to foods to which nanotechnologies have been applied. In cases where risk assessment is not providing conclusive proof of safety, market access would be prevented until sufficient proof of safety is demonstrated.

3.3.5 Food Safety Objectives
3.3.5.1 Safety Limits (5)
In many systems limits are set on the presence in foods of toxic substances and pathogenic organisms. If it would be possible to connect hazards to certain concentrations of ENPs – in general or of a certain kind – regulators could accordingly set limits to such concentrations.

3.3.6 Banned Contamination
3.3.6.1 Zero Tolerance (6)
A specific limit is the limit zero. It is applied to substances considered the most hazardous. Zero is specific because in practice it means the limit of detection, which in turn is a measure with a meaning that develops with technological progress in the field of detection.

In this sense 'zero' is an open and responsive norm whose meaning adapts itself to the advance of science. In this particular case, however, the risk exists that the adaptation of the meaning of the norm to growing understanding of the concept decreases instead of increases businesses possibilities to comply with the norm.

3.4 PROCESS

3.4.1 Hygiene
If nanotechnology is applied not with the objective to have ENPs in the end product, but to achieve some objective in the production, then

food hygiene is the context to manage possible risks. It seems desirable to have an obligation in place to fully apply hazard analysis and critical control points (HACCP)[10] to all processing relating to nanofoods. HACCP requires businesses to keep themselves up to date with regard to the potential hazards associated with — in this case — the use of nanotechnology and to continuously ensure that these hazards are under control.

3.4.2 Monitoring

To the extent that the presence of ENPs in food cannot be detected, monitoring of processes is needed in order to make their presence known. Imposing upon businesses the obligation to monitor and document the application of nanotechnology and to pass on this information down the food chain seems indispensable.

3.4.3 Traceability

On the one hand, traceability is the logical sequel to monitoring of the process within the business, i.e. keeping the information on the presence on ENPs available throughout the food chain. On the other hand traceability is an instrument to ensure safety by enabling to identify causes of food safety incidents and to remove consequences.

3.5 COMMUNICATION

Pre-packaged food shall not be described or presented on any label or in any labelling in a manner that is false, misleading or deceptive or is likely to create an erroneous impression regarding its character in any respect.[11]

3.5.1 Identity of Producer

To trace back a product to its producer for reasons of liability, enforcement or risk management, it is important that the label provides information on the identity of the producer or of another business that takes full responsibility for the product.

---

[10]Or a similar system such as the hazard analysis and preventive control plan (HAPCP) system included in the US Federal Food Drug and Cosmetic Act (section 418) by the Food Safety Modernization Act (FSMA). For a synopsis of the FSMA, see Fortin, 2011.

[11]Article 3.1, General Standard for the Labelling of Prepackaged Foods, CODEX STAN 1-1985 (Rev. 1-1991), lastly amended in 2005. See also Article 31 of the FAO model food law.

The Codex Alimentarius requires the following to appear on the label:[12]

The name and address of the manufacturer, packer, distributor, importer, exporter or vendor of the food shall be declared.

3.5.2 Presence of Nanotechnologies

Is it desirable to require businesses to inform consumers about the (possible) presence of ENPs or the use of nanotechnology in the process? The regulatory framework should ensure that only safe products are placed on the market. In this sense nano-labelling to warn the consumer of risks associated with approved uses of nanotechnology in general should not be necessary. If it would turn out, however, that certain nanofoods can be considered safe in general, but present hazards to certain susceptible consumers, provision of information may be necessary. As a complement to other requirements regarding the use of nanotechnologies, such as monitoring and traceability, indication on the label of the use of nanotechnologies or the possible presence of ENPs on the label may be useful in particular as long as detection is problematic. If labelling requirements address nanotechnologies or foods to which nanotechnologies have been applied, it will be necessary to have clear definitions to be able to delineate the scope and meaning of such an obligation.

3.5.3 Absence of Nanotechnologies

In case businesses wish to declare on the label the absence of nanoparticles, or ENPs, in order to avoid misleading the consumer it is important to have a clear concept of what such declaration means. A standard defining ENPs and their absence will be helpful.

3.6 FOOD CONTACT MATERIALS

Important applications of nanotechnologies take place in the packaging industry. The interaction between package (and other FCMs) and foods may result in the transfer of nano-engineered materials or ENPs to the food. The regulatory system should ensure that this does not lead to unacceptable risks to consumers.

---

[12]Article 4.4, General Standard for the Labelling of Prepackaged Foods, CODEX STAN 1-1985 (Rev. 1-1991), lastly amended in 2005.

In the USA packages and other FCMs are regulated as part of the food (indirect additives). In the EU they have separate legislation. Novel packaging materials are subject to premarket approval.[13] Active substances that may affect the food need to be approved as additives. Either way, the possible impact of FCMs on food safety should be covered by the regulatory system. Given the similarity in approach here we will limit our discussion to the approval of additives and other foods. The same argument applies, *mutatis mutandis*, to FCMs.

3.7 RECOMMENDATION

The above leads us to recommend the following elements for a food regulatory system. To ensure the safety of foods to which nanotechnologies have been applied, the regulatory system should ensure that authorities have information regarding the identity of businesses that bring such foods to the market. These businesses must be responsible and liable in general for compliance with the regulatory requirements and for food safety in particular.

Of the different approaches to food safety, we believe case-by-case premarket approval on the basis of risk analysis to be the most suitable for foods to which nanotechnology has been applied. We will elaborate this in Chapter 4.

As long as methods of detection fall short, information on the presence of nanomaterials should be ensured through monitoring, tracing and labelling.

Nano-specific requirements apply to foods containing nanomaterials. Nanomaterials are materials resulting from the application of nanotechnology. As elaborated in Appendix B and based on the literature review therein, we propose for this purpose to define nanotechnology as follows:

1. *The reductive and/or integrative activity to matter by which structures, systems or devices created at the nano-micrometre scale, which contain properties that are new in the sense that these properties would be absent if such activity would not have been implemented.*

---

[13]Regulation (EC) 1935/2004.

2. *Unless proven otherwise, applications of structures, systems or devices with a dimension below 300 nm are presumed to possess such properties.*

As explained (see Appendix B), the second part of the definition is of special importance in a system of responsive regulation in case of potentially dangerous foods, where the characteristic of these potential dangers may become apparent in the future.

CHAPTER 4

Case-by-Case Premarket Approval

4.1 INTRODUCTION

The system of choice to deal with emerging risks is to require favourable risk assessment as the condition for market entry. Often such a system applies so-called positive lists. This means that all foods and food ingredients falling within a certain category are banned, except those that have been included in the list of products that may be used. Inclusion in the list is based on risk assessment.

In adapting existing approval schemes or setting up new ones, several choices need to be made. In this chapter we touch upon five of these: the object of the approval scheme in Section 4.2, the addressee of approval in Section 4.3, the scope of approval in Section 4.4, the assessment needed for approval in Section 4.5 and a possibility to escape the approval requirements in the case that a substance is known to be safe in Section 4.6. The last section in this chapter (Section 4.7) takes the stock of the findings in terms of the adaptive qualities of the proposed scheme.

4.2 OBJECT OF APPROVAL

To what does the approval requirement apply? We find three types of criteria used to identify products that are submitted for premarket approval: products of a certain *origin* (e.g. genetically modified foods in the EU), products used for a certain *purpose* (e.g. food additives in the Codex Alimentarius, in Australia/New Zealand and in the EU) and *new* foods regardless of what makes them new and for what they are used (e.g. novel foods in the EU, Australia and New Zealand,[1] food additives in the USA[2]).

If we want foods to which nanotechnologies have been applied to undergo premarket approval, the question presents itself whether by analogy to genetically modified foods a specific category of nanofoods should be defined and submitted to an approval requirement, or whether such foods can fit into a wider category such as the categories of food additives or novel foods.

4.2.1 Nanofoods

The advantage of a separate approval procedure for foods to which nanotechnologies have been applied is that the procedure can be tailored to the specificities of nanotechnologies. However, if a separate approval procedure is opened for foods to which nanotechnologies have been applied, it is necessary to define which foods are within the scope of such procedure. This need does not arise if nanofoods fall within an already existing category of foods to which an approval requirement applies or if they can be brought within such a category by adapting its scope.

4.2.2 Additives

The Codex Alimentarius defines a food additive as: *any substance not normally consumed as a food by itself and not normally used as a typical ingredient of the food, whether or not it has nutritive value, the intentional addition of which to food for a technological (including organoleptic) purpose in the manufacture, processing, preparation, treatment,*

---

[1]According to FSANZ standard 1.5.1 novel foods are non-traditional food this means:
a. 'a food that does not have a history of human consumption in Australia or New Zealand; or
b. a substance derived from a food, where that substance does not have a history of human consumption in Australia or New Zealand other than as a component of that food; or
c. any other substance, where that substance, or the source from which it is derived, does not have a history of human consumption as a food in Australia or New Zealand.'
[2]Section 201(s) FFDCA [21 USC 321]: 'The term "food additive" means any substance the intended use of which results or may reasonably be expected to result, directly or indirectly, in its becoming a component or otherwise affecting the characteristics of any food (including any substance intended for use in producing, manufacturing, packing, processing, preparing, treating, packaging, transporting or holding food; and including any source of radiation intended for any such use), if such substance is not generally recognized, among experts qualified by scientific training and experience to evaluate its safety, as having been adequately shown through scientific procedures (or, in the case of a substance used in food prior to January 1, 1958, through either scientific procedures or experience based on common use in food) to be safe under the conditions of its intended use (...)'.

packing, packaging, transport or holding of such food results, or may be reasonably expected to result (directly or indirectly), in it or its by-products becoming a component of or otherwise affecting the characteristics of such foods. The term does not include contaminants or substances added to food for maintaining or improving nutritional qualities.[3] Three elements stand out in this definition: a substance (a) not normally consumed as a food (ingredient), (b) added to a food and (c) for a technological purpose.[4]

Only food additives listed in the Codex General Standard for Food Additives are recognised (by Codex) as suitable for use in foods. Only food additives that have been assigned an ADI or have been determined, on the basis of other criteria, to be safe by JECFA and an International Numbering System (INS) designation by Codex are considered for inclusion in that Standard.[5]

Obviously foods to which nanotechnologies have been applied may fall within the ambit of this concept of additives. In that situation it is important as to whether nanoadditives are qualified as a separate category of additives and as such need separate approval. However, as nanotechnologies are not applied exclusively to produce food components that exercise a technological function within the food, the category of food additives can never encompass all relevant applications of nanotechnologies. For this reason, the question if foods to which nanotechnologies have been applied fit into a more general category will always remain relevant. If food additives are lex specialis to this more general category, nanofoods fulfilling the definition of food additives will fall back into the scope of food additives when they meet the definition.

4.2.3 Novel Foods

The Codex Alimentarius does not apply a concept 'novel food'. In several jurisdictions – USA (food additives), Australia, Canada, EU, and Japan – a requirement is applied for foods that are new to that jurisdiction to undergo an approval procedure before market access is granted.

---

[3]Article 2(a) Codex General Standard for Food Additives, CODEX STAN 192-1995 (lastly revised 2009).
[4]The US concept of additive quoted in footnote 2 differs from the Codex concept of additive in that the elements (a) and (c) seem to be absent.
[5]Article 1(1) Codex General Standard for Food Additives, CODEX STAN 192-1995 (lastly revised 2009).

If foods to which nanotechnologies have been applied are not submitted to a separate approval procedure, but judged in the context of such a general approval procedure, the issue to define such foods shifts to the question of how to decide if they are new as understood in the context of the said framework.

As nanotechnologies are applied to create substances with new properties, it seems likely that in many instances the criterion of newness will be met. If necessary in a certain context, measures can be taken to ensure that the applied general criterion of newness encompasses the food application of nanotechnology.

In Chapter 5 we will discuss how to bring nanofoods within the ambit of approval requirements in situations where it is considered doubtful as to whether they are 'novel' by definition.

4.3 SUBJECT OF APPROVAL

To whom does the approval grant rights? Approval schemes come in two different guises. Some schemes are specific in that they address the applicant. Others are generic in that they address the product. In a generic scheme, the product is placed on a (positive) list. All businesses are allowed to bring the approved product to the market. In a specific scheme, the approval decision addresses the applicant, authorising the applicant to bring the product at issue to the market. All others who would want to bring an identical product to the market would need approval as well. A specific scheme rewards the applicant for the investment made in the safety assessment and in the procedure by granting the applicant an exclusive right. The downside is that repeated risk assessment is required even though the outcome of the risk assessment is already known to the authorities (i.e. the product is safe). The downside to a generic scheme is that it rewards the second to come to the market. This second benefits from the investment made by the applicant, without having made the effort.

Currently the EU is experimenting with an in-between form: data protection. If the application is based on proprietary data, for a certain period this data can be used only to the benefit of the applicant. Only after the data protection period has elapsed, the approval becomes truly generic. A second applicant, who wishes to enter the market before that time, will have to provide its own data. Data protection rewards the investment made in scientific research, not in the approval procedure as such.[6] An easier compromise between the interests served by generic approval and specific approvals would be to grant the applicant an exclusive right for a limited period of time only.

4.4 SCOPE OF APPROVAL

Which uses of the product are under approval? Many raw materials and products are used as food for humans as well as feed for animals. Markets for food and feed are largely separate but related. Experience has shown that many food safety issues started in feed fed to food producing animals. For this reason, the safety of feed may be relevant not only for the health of animals, but for the health of consumers as well. The StarLink case[7] has shown that practically speaking it is impossible to ensure that raw materials approved for use in feed only, do not end up in the food chain for humans. In the EU system for genetically modified foods the consequence of this has been drawn in that a product that can be used both as a feed and as a food can only be approved for both uses or not at all.[8] Approval only for feed is not possible. The logic is apparent and seems to make sense in the context of nano-technologies as well.

4.5 ASSESSMENT

A decided advantage of a premarket approval procedure on a case-by-case basis is that all the uncertainties that currently surround food applications of nanotechnology can be given a place. The applicant has an interest to actively contribute to solving problems. The business sponsoring a specific nanofood must present an application scrupulously identifying the product including methods of detection. Further, the procedure needs to have methods of risk assessment that, applying state of the art science, can indicate the likelihood that the product will cause adverse effects to consumers.

Most premarket approval schemes only apply negative criteria, that is to say they focus on risks and do not take into account benefits. One of the current debates on risk assessment raises the issue of whether

---

[6]For a discussion of other problematic aspects of the EU data protection scheme, see Carlson et al., 2010a,b.
[7]See Segarra and Rawson, 2001.
[8]See van der Meulen, 2007.

benefits can be accepted to outweigh certain risks such as (potential) allergenic properties.[9] In pharmaceutical law, for example, usually it is accepted that a beneficial medicine may have certain side effects. In foods, usually side effects are not accepted.

4.6 GENERALLY RECOGNIZED AS SAFE

A concern in the EU system of premarket approval for novel foods is that no exception exists to the requirements that foods that were not on the market before the cut-off date (15 May 1997) must be explicitly approved to be accepted to the market. There is no possibility included in the framework to exclude certain categories from this requirement if it has been sufficiently established that the category as such does not pose a relevant risk. If it is established that foods containing engineered nanoparticles are novel in the sense of the regulation this is true for each new foods containing such particles. If over time science would establish that the persistence of particles is a relevant risk factor,[10] each new food containing engineered non-persistent nanoparticles that have not yet been approved would still have to undergo safety assessment. In other words, in the EU system there is no way out of novelty. In this sense the regulatory approach included in the EU Novel Foods Regulation is not responsive. The American system holds a solution to this problem in the concept of 'GRAS'. A food additive (in the wide sense the Americans give to this word) that has not been on the US market before 1958 has to undergo safety assessment, except when it is 'Generally Recognized as Safe' (GRAS). In other words if within the scientific community consensus has been reached on the safety of a certain product (category), it no longer needs to be risk assessed by authorities.[11] This system is responsive in that scientific insights in society translate into the meaning of the law without the need to change the text of the law.

4.7 A RESPONSIVE SYSTEM?

Experience in the EU shows that the responsive capacities of a regulatory system largely depend on the way it is applied. In 15 years of

---

[9]Safe Foods <http://www.safefoods.nl/en/safefoods.htm>; EFSA, 6th Scientific Colloquium Report, Risk-benefit analysis of foods: methods and approaches.
[10]Or the number of dimensions at nanoscale.
[11]On a voluntary basis FDA's opinion can be sought as to the GRAS status of a product.

application of the Novel Foods Regulation, the EC has been scant in providing its reasoning in its decisions. By consequence vital concepts that were vague at the outset, have not acquired a body of precedents and are still vague 15 years after.[12] The Novel Foods Regulation could work like a principles-based responsive regulatory system, but only if it would be applied in a transparent discursive manner. What would be needed, for example, for the elusive concept of 'novel food' to acquire substance, is a publicly available reasoning in each case where the responsible public authority (i.e. the EC), responding to positions taken by interested parties, makes explicit why it considers the product at issue to qualify as a novel food.

The concept of premarket approval of novel foods holds the potential to adapt to food applications of nanotechnology. For this potential to come to the fore, it needs to be applied in a conscientious way.

[12]See van der Meulen, 2009c.

CHAPTER 5

Nano-Specific Regulation

5.1 TOWARDS A LEGAL CONCEPT

5.1.1 Market Access

As we have shown, safety of foods to which nanotechnologies have been applied *can* be regulated without a legal definition of nanofoods. By consequence, the difficulty in specifying what exactly nanotechnology is and what gives a food relevant 'nano' properties does not constitute an insurmountable obstacle to their regulation.

Nevertheless, in so far as it is necessary or desirable to create nanofood-specific provisions, a concept or definition of such foods is needed. Such concept should serve the objective of the regulation that is to regulate health risks that a substance may pose due to new properties achieved through the applications of nanotechnology.

From the point of view of health risks, the core issue is the *properties* of the substances, rather than the *way* these properties have been brought about: nanotechnologies. Therefore a regulatory concept should somehow catch these properties. As we are now moving beyond the scope of principles-based general food safety law, some link with the technology should be established as well. One of the first provisions,[1] maybe *the* first, in food legislation explicitly dealing with nanotechnology, is the EU Regulation on food additives. Article 12 of Regulation (EC) 1333/2008, reads:

> **Changes in the production process or starting materials of a food additive already included in a Community list**
> When a food additive is already included in a Community list and there is a significant change in its production methods or in the starting materials used, or there is a change in particle size, for example through nanotechnology, the food additive prepared by those new methods or materials shall be considered as a different additive and a new entry in the

[1]The EU attempted a recast of the Novel Foods Regulation. It included two approaches to regulating nanotechnology. The proposed nano-specific provisions are included in Appendix D and may provide some inspiration. The EC's proposal was rejected by the European Parliament. The disagreement between the institutions did not include the approach to nanotechnology, so it seems likely that a new proposal will be similar at this point.

Community lists or a change in the specifications shall be required before it can be placed on the market.

The approach chosen in this provision is principles based and adaptive in that it uses open legal concepts such as 'particle size' and 'nanotechnology' without giving a strictly delineated legal definition, but instead leaving it to the discourse between risk assessors, businesses and public authorities to give them meaning. This approach seems suitable to bring food applications of nanotechnology within the ambit of existing premarket approval schemes. If we slightly rephrase this quote to fit to other schemes as well, the text would run something like this:

When a food is already legally on the market and there is a significant change in its production methods or in the starting materials used, or there is a change in particle size, for example through nanotechnology, the food prepared by those new methods or materials shall be considered as new and a new entry in the [applicable positive list] shall be required before it can be placed on the market.

Without a definition of nanotechnologies this provision is vague. However, its reference to significant change in methods and materials makes the principle involved sufficiently clear for application and development in practice, for the purpose of premarket approval.

5.1.2 Labelling and Traceability

For labelling and traceability purposes, probably a clearer cut delineation is unavoidable.

Nanotechnologies as applied in the context of the food chain may lead to synthesis of new substances and to presence of engineered particles and constructs of nanosize. Substances should be detectable by conventional means and mentioned on the label as ingredients or additives. Critical are the ENPs. For the reasons set out in Section 3.4, traceability of foods containing ENPs should be ensured and the presence of such particles should be labelled. This labelling requirement is more for the benefit of inspection than for consumer information.

Appendix B explains more in detail how we attempt to come to a delineation of the concepts of nanotechnology and nanofood. For the purpose of labelling, we propose the following formula.

The presence of products resulting from processes meant to create particles or constructs with at least one dimension below 300 nm should be mentioned on the label.

Often in attempts to define nanotechnology a cut-off is proposed at 100 nm.[2] The here proposed provision takes a margin by opting for 300 nm. By doing so, on the one hand it stays away from scientific discussions about definition of nanotechnology. If, on the other hand, it is true that relevant properties are achieved below 100 nm dimensions only a cut-off at 300 nm will not create problems for compliance as there will be no intentional construction between 100 and 300 nm.[3]

---

[2]See Appendix B.

[3]In the EU a different approach has been chosen. Regulation (EU) 1169/2011 on food information to consumers states (in Article 18(3)): *All ingredients present in the form of engineered nanomaterials shall be clearly indicated in the list of ingredients. The names of such ingredients shall be followed by the word 'nano' in brackets.* 'Engineered nanomaterial' in turn is defined (in Article 2 (2)(t)) as: *any intentionally produced material that has one or more dimensions of the order of 100 nm or less or that is composed of discrete functional parts, either internally or at the surface, many of which have one or more dimensions of the order of 100 nm or less, including structures, agglomerates or aggregates, which may have a size above the order of 100 nm but retain properties that are characteristic of the nanoscale. Properties that are characteristic of the nanoscale include:*
i. *those related to the large specific surface area of the materials considered; and/or*
ii. *specific physico-chemical properties that are different from those of the non-nanoform of the same material.*

CHAPTER 6

Regulatory Burdens

6.1 BURDENS

One might argue against the regulatory structure we propose in this study on the basis of the burdens it will place on innovation and businesses. Indeed the regulatory burden should not be underestimated. Approval procedures in particular are lengthy and costly and often their outcome is uncertain.[1] By consequence premarket approval schemes are seen as a serious barrier to innovation and trade. Only the strongest companies can afford to submit a product for approval. Thus premarket approval schemes favour large companies over smaller companies and companies from the first world over companies from the third world. In practice companies are observed to avoid approval schemes by not engaging in innovation, by hiding innovation from authorities, by selecting markets without (effective) approval requirements and bypassing markets where safety needs to be proven scientifically. By consequence, premarket approval schemes may not always provide the desired level of safety protection to all consumers.

6.2 OTHER CONCERNS

Regulatory burdens are not the only concern the proposed system, and in particular the requirement of premarket approval, raises. Another is WTO compatibility. Premarket approval of several types of products seems to be common among WTO members. These types of products include pharmaceutical products, plant protection products and food additives. Nevertheless, some WTO members have informally contested such premarket approval requirements within the WTO SPS Committee. They argue that the SPS Agreement places the burden of proof on the states that are members of WTO and that the SPS Agreement does not allow for a reversal of the burden of proof such as inherent in premarket approval schemes where the burden is on businesses. So far no formal dispute settlement has resolved this issue. Nevertheless, the case for premarket approval would be stronger if it

---

[1] On this topic see van der Meulen, 2009c.

were agreed upon internationally or if authorities setting the trade bar-rier would provide support in dealing with it.[2]

6.3 ALLEVIATING BURDENS

It can be argued that it is fair to leave the burden to prove safety with the sponsors of innovative foods. The value of such argument seems limited, however, if the practical effects are counterproductive in the sense that innovation is delayed or shifted to countries where consu-mers enjoy limited protection. It would be better to limit the burdens as far as possible. Measures that can be taken to this effect include the following. First, adoption of a GRAS-like exemption from approval requirement that enables the competent authority to lift the approval requirement for certain products or processes regarding which suffi-cient data are available for a general conclusion regarding absence of risk.[3] Second, avoid requiring scientific proof in situations where the safety of a food is not uncertain because conclusive risk assessment is already available. This argues against an exclusive system and in favour of a generic system.[4] Third, active participation of the public sector in risk assessment and serious efforts on the part of authorities to deal with the procedure in an efficient and effective manner, creat-ing as little additional costs in time and finance for the applicants as possible. Fourth, mutual recognition of risk assessment.[5] Fifth, strict compliance by public authorities with the obligations applying to them such as, in particular, respecting deadlines for taking decisions. Finally, authorities can do much in terms of compliance assistance. That is to inform businesses willing to comply with their obligations on how to do it and support them in achieving it.

6.4 GLOBAL HARMONISATION?

The analysis in this study is largely based on systems found in national (including EU) systems of food law. The question presents itself how-ever, what is the most appropriate level of regulation. Premarket approval schemes currently exist in the USA, Canada, the EU, Japan, Australia and New Zealand. So far these jurisdictions fare poorly in

---

[2]On this topic, see also Szajkowska, 2012, in particular Chapter 3: Science-based regulation?
[3]See Section 4.6.
[4]See Section 4.3.
[5]See Section 6.4.

mutually recognising each other's risk assessment. The consequence is that innovative businesses suffer the administrative burden of multiple approval procedures. Apparently, risk managers do not recognise each other's risk assessments,[6] let alone each other's risk management decisions. It would be much to their credit if they would manage to terminate this situation. On top of this, many countries do not have the capacity or infrastructure for safety assessment and approval. The level of safety assurance for consumers differs considerably between countries.

The Joint FAO/WHO Food Standards Programme has an infrastructure in place for risk assessment at the global level. It would be to the benefit of the whole world — authorities, consumers and businesses alike — if case-by-case assessment and approval of nanofoods would be concentrated at the global level to be applied by countries that do not have an approval system of their own and to be supported and recognised by countries that do.

---

[6]The closest thing to an exception seems to be that in the USA the GRAS status of a product can probably be based on risk assessment by JECFA or EFSA.

CHAPTER 7

A Responsive Regulatory System

Regulatory systems lay down rules to be applied to situations that to a large extent are still in the future at the time of rulemaking. Thus, inevitably the system will apply concepts that have a certain vagueness and require interpretation in their application to cases that present themselves. This is true in general; it is certainly true for the approach proposed in this study. A system is responsive to the extent it is capable to increasingly give meaning to the vague notions in the law. In other words a formal (common law) or informal (civil law) system of precedent is needed not only in litigation but in administrative practice as well. Key features to this end are transparency and accountability. Transparency means that the content of the learning process is visible for all stakeholders concerned; accountability means that stakeholders applying the systems lay themselves open to scrutiny.

To be a responsive system, the system must be embedded in a context that invites rather than discourages the free exchange of opinions in procedures of approval, review and litigation. Public authorities – be they administrative or judiciary – must take responsibility to set meaningful precedents by providing ample motivation for each application and interpretation of the system. Why do the authorities deem the system applicable – or not – in a given case?[1] Why do they consider the applied method of risk assessment suitable – or not – as a basis for risk management decisions? Why do they consider the risk identified to be acceptable – or not – from the perspective of consumer protection?

Exchange of views is vital. In the preparation stage of decisions, stakeholders' views should be considered. Ideally decisions should be open to contestation in a procedure of full review. All decisions should give full account of the reasons and considerations on which they are based. As little barriers as possible should hamper stakeholders from

---

[1]Experience in the EU shows that the responsive capacities of a regulatory system largely depend on the way it is applied. In 10 years of application of the Novel Foods Regulation, the EC has been scant in providing its reasoning in its decisions. By consequence vital concepts that were vague at the outside have not acquired a body of precedents and are still vague 10 years later.

bringing issues to a court of law. As science has become part of the law, discussion of the scientific merits of the case should be part of the review by the courts and not left to the discretion of the administration. The system of dispute settlement under the WTO provides a good example[2] where science can be pitted against science to be judged on the merits of science.

---

[2]EU case law by contrast provides an example where the way risk assessment is applied in risk management is left to the discretion of the administration (i.e. the EC). See for example, Case T-326/07, *Cheminova* v. *Commission*, [2009] ECR 0000, at 106.

CHAPTER 8

Conclusions and Way Forward

Ongoing developments in nanoscience and application of nanotechnologies in the food chain present a challenge to regulatory systems protecting food safety.

8.1 RECOMMENDATIONS TO LEGISLATURES

On the basis of the arguments and observations presented previously, we recommend that regulatory frameworks be checked and reviewed for presence of the following elements and − in case of absence − to consider inclusion.

1. A general basis holding businesses that bring food products to the market responsible and liable for the safety of these foods.
2. A registration obligation encompassing all businesses that bring foods to the market to which nanotechnology has been applied.
3. A premarket approval requirement on a case-by-case basis for new foods including food applications of nanotechnology that sets a favourable outcome on science-based risk assessment as the condition to market access.
4. An obligation to fully implement HACCP for all businesses applying nanotechnologies in the food chain or dealing with food applications of nanotechnology.
5. An obligation to document and trace the possible presence of ENPs in food products.
6. An obligation to mention on the label of foods to which nanotechnology has been applied
   a. The identity and address of the responsible business
   b. The possible presence of ENPs.

8.2 RECOMMENDATION TO THE SCIENTIFIC COMMUNITY

Science must proceed in identifying risks and methods of detection and risk assessment.

8.3 RECOMMENDATION TO INTERNATIONAL RISK ASSESSMENT AUTHORITIES

FAO, WHO, OECD, FDA, EFSA and other key players should take responsibility to come to a worldwide system of (mutual) recognition of risk assessment, instead of the current situation where procedures must be repeated in several countries.

APPENDIX A

Analytical Framework

A.1 ELEMENTS OF FOOD SAFETY REGULATION

A.1.1 Introduction

As indicated in Section 2.1, when in this study we refer to 'regulatory framework', this includes the entire hierarchy of hard and soft law: of principles, acts, regulations, guidelines, codes of conduct, implementing policies, scientific policies, conformity assessment requirements, powers of inspection, sanitation and sanctioning and any other tools (by whatever name) that are in place to deal with food safety issues.[1] The regulatory framework encompasses provisions of a general nature empowering public authorities and addressing food businesses and moves down to include administrative decisions regarding the marketability of categories of foods and the compliant status of specific foods on the market. In this appendix we present the frame of analysis that we apply to food safety regulation in general and that we have also taken as reference for the analysis in the study of the regulation of foods to which nanotechnology has been applied.

A.1.2 Frame of Analysis

The frame of analysis is based on a comparative study of food law as it has developed in different parts of the world. Food regulatory systems around the globe differ greatly in appearance of rules and requirements originating from different authorities. Their content, however, usually addresses similar issues.[2] These issues are depicted in Figure A.4. This figure comprises several notions that are introduced in Figures A.1–A.3. First of all, food safety regulation embodies a

---

[1]See also, the joint FAO/WHO publication (2003). Assuring Food Safety and Quality: Guidelines for Strengthening National Food Control Systems; Vapnek and Spreij, 2005.
[2]Or so we believe. Comparative food law is still in its infancy. See for example Boisrobert et al., 2009.

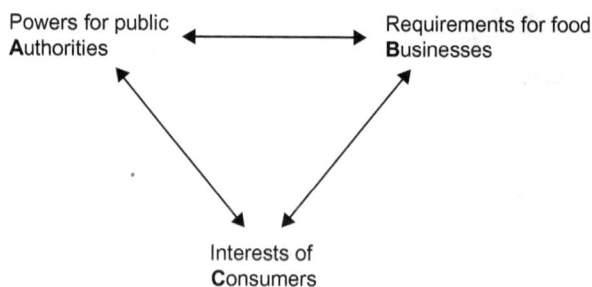

Figure A.1 The triangular relations underlying the structure of food law.

triangular relation between consumers, public authorities and businesses. Consumers need rights to hold businesses liable and to demand action from authorities. Such rights need to be embedded in a context of transparency. Public authorities need instruments ('powers') to regulate and enforce. Businesses are responsible for the safety of their products.

At the basis of Figure A.4 are the interests of consumers that are the purpose of the entire framework. At the top are the general concepts and principles from which the framework is derived. In this study the principles of risk analysis, precaution, business responsibility and liability are elaborated.

The left-hand side of Figure A.4 shows the rules addressing public authorities. These encompass requirements on standard setting and on administrative action such as approval and banning of products, enforcement and incident management. Activities of public authorities in food safety regulation are often grouped under the heading 'risk management'. Risk management encompasses the setting of general rules applying to businesses (and thus represented at the right-hand side in the figure) and measures taken with regard to food-related issues such as incident management and enforcement. Incident management is at issue if *foods* are not in compliance with safety requirements. Enforcement is at issue if *people or businesses* do not comply with their obligations. To a large extent these categories overlap, but they are not identical. Risk management is an element of risk analysis also including risk assessment and risk communication as prerequisites of risk management. Chapter 2 of the study addresses requirements for public authorities on giving substance to the regulatory framework and on administrative powers. For this reason no further elaboration of this part of the framework is needed here.

Powers for public
Authorities

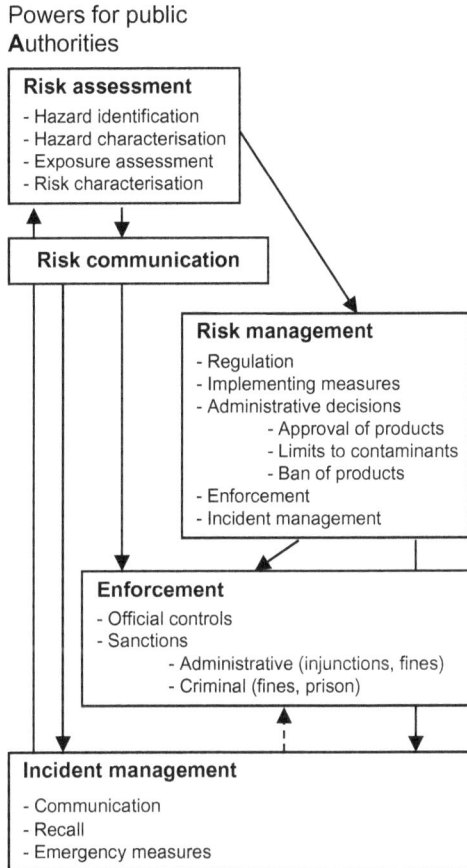

Figure A.2 Public powers within the frame of analysis.

The right-hand side of Figure A.4 shows what is probably the most important for the purpose of regulating the safety of nanofoods: the requirements on businesses ('what' do they have to do or refrain from, 'what' do they have to achieve). The right-hand side is elaborated here below. Within the regulatory framework addressing businesses, we distinguish four focus areas. The first being requirements on the business itself including its premises and equipment, the second and most important: requirements on the product (the food as such), the third: requirements on the process of production and trade, and the fourth: requirements on the communication about the food in labelling and advertisement.[3]

Requirements for food
Businesses

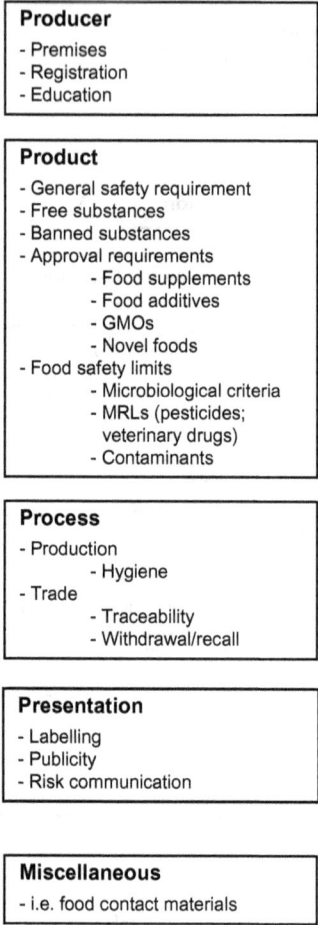

Producer
- Premises
- Registration
- Education

Product
- General safety requirement
- Free substances
- Banned substances
- Approval requirements
    - Food supplements
    - Food additives
    - GMOs
    - Novel foods
- Food safety limits
    - Microbiological criteria
    - MRLs (pesticides;
      veterinary drugs)
    - Contaminants

Process
- Production
    - Hygiene
- Trade
    - Traceability
    - Withdrawal/recall

Presentation
- Labelling
- Publicity
- Risk communication

Miscellaneous
- i.e. food contact materials

Figure A.3 Requirements for businesses within the framework of analysis.

The middle of Figure A.4 shows a different dimension: conformity assessment ('how' is it established if the 'what' has been met).[4] This

---

[3]All the rest is grouped in Figure A.4 under the fifth heading 'miscellaneous'.
[4]Or, in the words of ISO, the International Organization for Standardisation: Conformity assessment is the name given to the processes that are used to demonstrate that a product (tangible) or a service or a management system or body meets specified requirements.

conformity assessment encompasses tools both for compliance (right hand) and for law enforcement (left hand). It includes topics such as documentation, sampling, risk assessment, certification and inspection. Ideally primary ('what') and secondary ('how') requirements are balanced in the sense that businesses can establish if they comply and authorities if they do not. Conformity assessment is not a separate topic in this study but plays a role with regard to all topics. With regard to nanofood, conformity assessment is problematic for as long as methods of detection and risk assessment are lacking.

Combining all the above in a figure, we present a framework for the analysis of food regulatory systems. It is shown in Figure A.4.

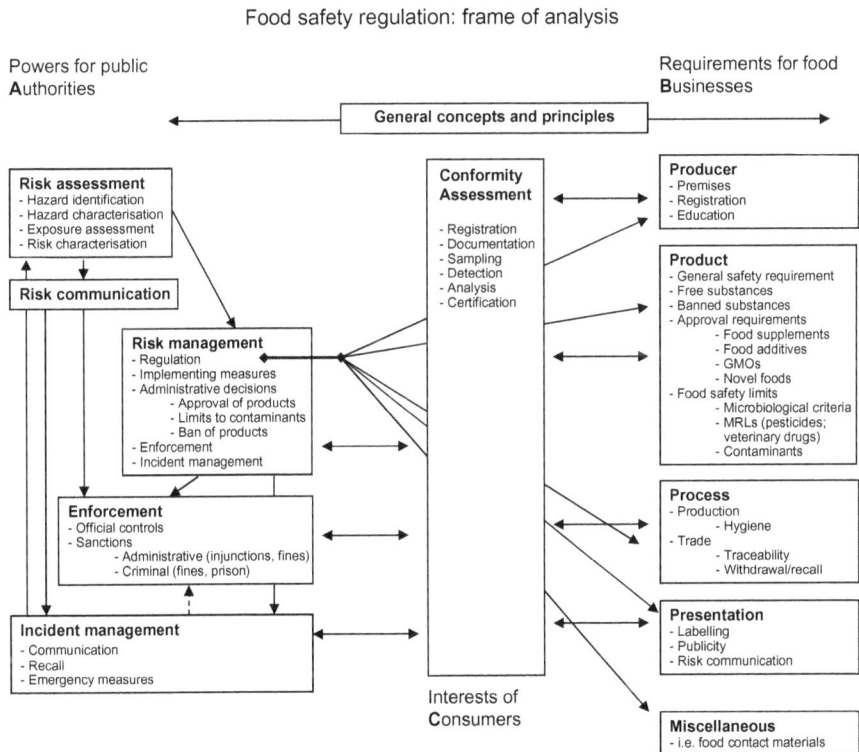

Food safety regulation: frame of analysis

Powers for public Authorities

Requirements for food Businesses

General concepts and principles

Risk assessment
- Hazard identification
- Hazard characterisation
- Exposure assessment
- Risk characterisation

Risk communication

Risk management
- Regulation
- Implementing measures
- Administrative decisions
   - Approval of products
   - Limits to contaminants
   - Ban of products
- Enforcement
- Incident management

Enforcement
- Official controls
- Sanctions
   - Administrative (injunctions, fines)
   - Criminal (fines, prison)

Incident management
- Communication
- Recall
- Emergency measures

Conformity Assessment
- Registration
- Documentation
- Sampling
- Detection
- Analysis
- Certification

Interests of Consumers

Producer
- Premises
- Registration
- Education

Product
- General safety requirement
- Free substances
- Banned substances
- Approval requirements
   - Food supplements
   - Food additives
   - GMOs
   - Novel foods
- Food safety limits
   - Microbiological criteria
   - MRLs (pesticides; veterinary drugs)
   - Contaminants

Process
- Production
   - Hygiene
- Trade
   - Traceability
   - Withdrawal/recall

Presentation
- Labelling
- Publicity
- Risk communication

Miscellaneous
- i.e. food contact materials

Figure A.4 Frame of analysis.[3]

[5]Elaborated on the basis of van der Meulen and van der Velde, 2008, 251. On the structure of food law see also van der Meulen, 2009b and van der Meulen, 2012.

A.2 REQUIREMENTS ON FOOD BUSINESSES

On the right-hand side, Figure A.4 shows the requirements for businesses: the first category addresses the business as such; the second addresses the food; the third the process of food production, trade and distribution; the fourth the communication about the food and at the end a category is reserved for other (miscellaneous) topics.

A.2.1 Requirements Regarding the Business

In some countries food businesses need to be registered.[6] Sometimes approval requirements exist for certain types of businesses or certain levels of education for staff. In hygiene legislation, requirements may be set to the layout and construction of premises.[7]

A.2.2 Product Requirements

Most modern food regulatory systems aim to protect health and safety of consumers. In these systems we may find six approaches to the product (the food as such), which can be subdivided into three categories.

A. *The product in concreto*
   1. Any food placed on the market must be fit for human consumption in terms of its safety.
B. *Categorical approaches to foods and food ingredients*
   2. Certain categories of foods are considered safe and can be used freely.
   3. Certain categories of foods are considered unsafe and banned.
   4. Certain categories of foods are presumed unsafe and therefore banned unless/until food businesses prove otherwise.
C. *Approaches to substances in foods*
   5. Public authorities set limits to the presence of substances or organisms they consider hazardous.
   6. Public authorities apply zero tolerance to the presence of substances or organisms they consider extremely hazardous.[8]

Different approaches will occur at the same time. The first (A) applies to the individual product and its effect in specific cases. A product

---

[6]The FAO Model Food Law provides for the following provision (Article 17):
   1. All premises including warehouses used for the preparation, sale, exposure or storage of food shall be constructed as prescribed.
   2. All food businesses shall apply for a licence in accordance with the prescribed procedures.

[7]Australia, see FSANZ Standard 3.2.3, EU see Regulation (EC) 853/2004.

[8]In practice one finds also zero tolerances for other reasons. These, however, seem hard to justify.

belonging to a category considered safe can still be unsafe due to its condition (decay or mishandling for example). The second group (B) applies to categories of foods and ingredients used in the production of food and the third (C) to categories of substances within foods.

A.2.3 Process

Regulators not only target the food product as such, but also the process by which it has been produced and traded. In this category of requirements we can think of topics like food hygiene and traceability. Food hygiene is both a means to achieve the objectives following from the requirements on the product and an end in itself in that it assures the control of hazards resulting from or related to the production process.

The Codex Alimentarius defines Food hygiene as: *'all conditions and measures necessary to ensure the safety and suitability of food at all stages of the food chain'*.[9] In particular HACCP (hazard analysis and critical control points) is a responsive system suitable to adapt advancing insights into the possibilities and hazards of nanotechnology. The HACCP system is elaborated in much detail in the Codex Alimentarius.[10]

According to the Codex, the HACCP system consists of the following seven principles:

1. Conduct a hazard analysis.
2. Determine the Critical Control Points (CCPs).
3. Establish critical limit(s).
4. Establish a system to monitor control of the CCP.
5. Establish the corrective action to be taken when monitoring indicates that a particular CCP is not under control.
6. Establish procedures for verification to confirm that the HACCP system is working effectively.
7. Establish documentation concerning all procedures and records appropriate to these principles and their application.

---

[9]Recommended International Code of Practice General Principles of Food Hygiene, CAC/RCP 1-1969, Rev. 4 (2003), p. 6.
[10]See Codex Alimentarius Commission. Food hygiene. Basic texts, fourth edition, WHO/FAO Rome 2009.

With regard to traceability, the FAO model food law holds the following provision (Article 18):

(1) *Food businesses shall establish and implement a system enabling them to identify any person:*
   (a) *who supplied; or*
   (b) *to whom they supplied;*

   a food-producing animal, food or substance intended to be or expected to be incorporated into a food.

(2) *Upon request of the Authority, food businesses shall make available all information collected under the system established under sub-Article (1).*

A.2.4 Communication

As of 20 September 2013, 167 states are party to the International Covenant on Civil and Political Rights (ICCPR). Article 19(2) and (3) ICCPR grants everybody freedom of expression to be limited only by law and only if necessary for the protection of high-ranking values such as rights of others and public health. Food legislation limits freedom of expression in that it requires certain information to be provided to consumers on the label of food products. It may make requirements on the voluntary provision of certain information and it may limit messages not desired by the authorities. The space to provide information on the label of food products is limited. It should only be required to provide additional information on the label if there is a good reason to do so like a need to know on the part of the consumer, or a strong desire to know in a relevant part of the consumer population.

APPENDIX *B*

On the Delineation of the Concept 'Nanotechnology'

For specifically regulating nanotechnology and its effects on food processing as well as on the safety of food and feed, it may be useful to set boundaries to the concept as a first orientation within a responsive system of nanotechnology food law. With setting boundaries, phenomena to which the rules and regulations are applicable are distinguished from those to which they do not apply. Also, risk assessment and risk management will have to be applied to concrete phenomena which can be ranked under the concept 'nanotechnology'. The notion that nanotechnology applies to materials with sizes less than 100 nm is not sufficient[1] and in some cases would lead to false conclusions about the range of scientific innovations to which the concept of nanotechnology refers. While the small size is indeed a characteristic of the playing field of nanotechnology (with two advantages: miniaturisation (i.e. increased relative surface area, which alters chemical reactivity[2] as well as novel quantum physics effects), a broader range should be adopted in which such phenomena are located: the continuum from nanoscale to microscale.[3] The term 'nanotechnology' was first proposed by Taniguchi,[4] referring to the manufacture of mechanical devices within the size range of nanometers.[5] Whatmore (1999)[6] states that 'the term has now expanded in its meaning to cover the fabrication and exploitation of entities (materials, devices etc.) in which there is a dimension of less than 100 nm, *which is critical to performance or behavior*' (italics by the authors). The US National Nanotechnology Initiative uses an

---

[1]Compare: Scientific Committee on Emerging and Newly Identified Health Risks (SCENIHR), 2008. The scientific aspects of the existing and proposed definitions relating to products of nanoscience and nanotechnologies. Directorate-General of Health and Consumer Protection/ European Commission, at p. 3.
[2]Hunt, 2004, 13.
[3]Bawa et al., 2005, 151.
[4]Taniguchi, 1974.
[5]Whatmore, 1999.
[6]Whatmore, 1999.

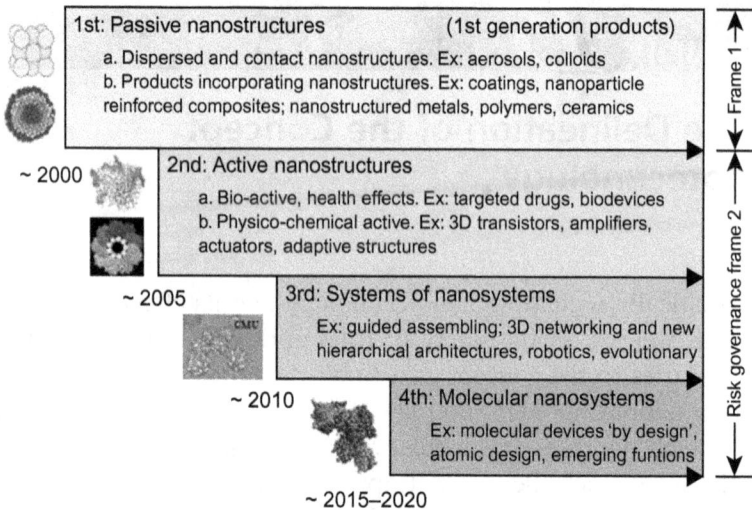

Figure B.1 *Applications of nanotechnology. http://www.crnano.org/whatis.htm.* Source: O. Renn (2006). Nanotechnology and the need for risk governance. Journal of nanoparticle research, 8(2).

explicit range in which nanotechnology can be identified:[7] 'Nanotechnology is the understanding and control of matter at dimensions between approximately 1 and 100 nanometers, where unique phenomena enable novel applications. Encompassing nanoscale science, engineering, and technology, nanotechnology involves imaging, measuring, modeling, and manipulating matter at this length scale'. Applications of nanotechnology spread fast (see[8] Figure B.1) and the ground-breaking novelty of the behaviour of nanoinventions urges for regulatory control.[9] However, 'as used today, the term nanotechnology usually refers to a broad collection of mostly disconnected fields'.[10] The range of applications is illustrated in Figure B.1, as well as the progress of nanotechnology. It becomes increasingly important in industry,[11] including food production and packaging.[12]

We conclude that the size or scale element, especially the delimitation to a certain range, may not be sufficient for setting effective

---

[7]<http://www.nano.gov/nanotech-101/what>.
[8]Center for Responsible Nanotechnology.
[9]Compare in this respect Mehta, 2004.
[10]Center for Responsible Nanotechnology.
[11]Siegrist et al., 2007, 459.
[12]See in this respect: Bouwmeester et al., 2009, 54 (table 1). Also, Sanguansri and Augustin, 2006, table 1.

boundaries to nanotechnology.[13] For instance, some of the novel properties of materials, systems and processes which are classified under nanotechnology occur even outside the border of 100 nm. Size reduction is only one side of the nanotechnology coin, as research is ongoing and applications are designed for the atomic element to larger structures and systems, which is known as a bottom-up approach. Any definition aims at expressing the essence of a concept. From the definitions which are available, essential elements which should be represented can be isolated to come to a definition which is not only recognised as a valid representation of a concept, but also a practical delineation of significant aspects. In our view, practical considerations should be included in delineating the concept, like the desire to control food safety. In this respect, the most important 'regulatory trigger'[14] being the uncertainty whether nanoforms of traditional products are really 'novel'. Or otherwise stated, ultimately it ends up with the question of whether existing 'substances'[15] in a rescaled form are *new* 'substances'.

Despite a persistent lack of knowledge concerning the aspects to include in the nanotechnology concept and the riskiness[16] of its application (expressed in toxicity $\times$ exposure levels),[17] delineation is desirable but also should be done with caution. A rigid delineation has as a disadvantage that future novelties would be excluded from the concept. This implies, that under conditions of technological progress a continuous process of adjustment of terminology and legal definitions could be expected, to prevent new applications from falling outside its boundaries. Present definitions often include conceptualisations as can be found in the Royal Academy and Academy of Engineering (2004),[18] addressing the concepts of nanotechnology and nanoscience: '***Nanoscience*** *is the study of phenomena and manipulation of materials at atomic, molecular and macromolecular scales, where properties differ significantly from those at larger scale.* ***Nanotechnology*** *is the design,*

---

[13]Compare ISO, 2008. Guideline TS 27687 at p. 1.

[14]Ludlow et al., 2007, 5.

[15]See in this respect SCENIHR, 2008, at p. 9. 'For the purposes of risk assessment in European Union chemicals regulation (1907/2006/EC), the term "substance" has a more precise and detailed definition than normally encountered. Such usage has to be taken into account'.

[16]Bowman and Hodge, 2007, 4.

[17]Zijverden and Sips, 2008.

[18]The Royal Society and The Royal Academy of Engineering, 2004. Nanoscience and nano-technologies: opportunities and uncertainties, July.

characterisation, production and applications of structures, devices and systems by controlling shape and size at the nanometre scale'. For instance, the Dutch research institute RIVM (2008)[19] defines nanotechnology as: *'the design, characterisation, production and applications of structures, devices and systems by controlling shape and size at the nanometre scale.'* As stated by Hunt, nanotechnology is 'the application of nanoscience in technological devices'.[20] The given delineation of the concept of nanotechnology has been adopted (in the same or quite similar terminology) by numerous research institutes, like for instance the Dutch RIVM or as a result of discussions with renowned researchers as in the following publicly available definition:[21] *'The design, characterization, production, and application of structures, devices, and systems by controlled manipulation of size and shape at the nanometre scale (atomic, molecular, and macromolecular scale) that produces structures, devices, and systems with at least one novel/superior characteristic or property'.*

The following basic elements can be noted:

- Size (or scale): atomic, molecular and macromolecular, however without the specification of a range like $1- < 100$ nm. As said, the 'differing properties' can also occur outside this range.
- differing properties (only in nanoscience here above, but in our view also applicable to nanotechnology).
- design, characterisation, production and applications of structures, devices and systems, saying that the use of nanotechnology leads to *novelties* – characteristics which are absent on a larger scale – which are causally related to the manipulation of size and scale.

It should be noted that the introduction of the term 'novel' also poses a problem, because (at least from a European viewpoint) this concept has been defined inadequately.[22] Size and scale applications can be reached bottom-up as well as top-down. The bottom-up approach refers to the design of structures, systems and devices for small elements to larger, while top-down is the systematic reduction of size and scale. Smaller size can lead to bigger surface area and induce improved water absorption, changes in flavour release, bioavailability

---

[19]Roszek et al., 2005.
[20]Hunt, 2004, 14.
[21]See <www.nanowerk.org>.
[22]See Chaudhry et al., 2008.

and faster rates of catalysis.[23] Applications in the food industry can occur in the fields of food ingredients, nanoparticle additives, packaging as well as water filtration, sensoring and traceability as a replacement or supplement to RFID.[24] Applications of nanotechnology are different depending on the scale (one-dimensional, two-dimensional or three-dimensional) of the structures, devices or systems which are created, as is illustrated in the following overview:[25]

Table B.1 Dimensions of Nanotechnology	
Nanoscale in one dimension	Monolayers: one atom or molecule deep
Two-dimensional nanomaterials	a. Carbon nanotubes b. Inorganic nanotubes c. Nanowires d. Biopolymers
Nanoscale in three dimensions	a. Nanoparticles, can be fixed or free; especially free nanoparticles can influence human health severely b. Fullerenes (Carbon60) c. Dendrimers d. Quantum dots
Source: Based on www.nanowerk.org.	

On the basis of the previous discussion the concept of 'nanotechnology' would have to include:

- the size/scale aspect (without any delineation to a certain range). Nanotechnology concerns the 'manipulation of matter at the atomic level'.[26]
- the outcome or result of application of nanoscience, where reduction significantly changes the properties of matter.
- the novelty of properties, 'enabling scientists to create specific molecular structures and devices'.[27]
- the causal relationship of novelty with manipulated size or scale.

Before presenting a functional definition of nanotechnology, we state the meaning of the concept as proposed by the ISO in its draft

[23]Sanguansri and Augustin, 2006.
[24]Chaudhry et al., 2008.
[25]Constructed on the basis of <www.nanowerk.com>.
[26]Bowman and Hodge, 2007, 3, referring to Drexler et al., 1991.
[27]Bowman and Hodge, 2007, referring to Forrest, 1989, 5.

business plan.[28] ISO states that 'nanotechnology' may be defined as either or both of the following:[29]

1. 'Understanding and control of matter and processes at the nanoscale, typically, but not exclusively, below 100 nanometres in one or more dimensions where the onset of size-dependent phenomena usually enables novel applications, where one nanometre is one thousand millionth of a metre;
2. Utilising the properties of nanoscale materials that differ from the properties of individual atoms, molecules, and bulk matter, to create improved materials, devices, and systems that exploit these new properties.'

It should be noted that the definition does not limit the nanoscale element to below 100 nm, which is of importance for the food industry, as 'many of the nanoscaled materials that are used consist of nanoscaled objects smaller than 100 nm which are dispersed in the product only in an aggregated or agglomerated state, such as micelle systems with a diameter of 300 nm'.[30] However, the concept 'not exclusively' provides a vague delimitation, which may be not functional within a legal context, as it does not assign a definite responsibility to proof safety of application to actors within the field of production and/or processing.

A generic delineation, providing a foothold for further research and juridical application, could look as stated below:

1. *The reductive and/or integrative activity to matter by which structures, systems or devices created at the nano-micrometre scale, which contain properties that are new in the sense that these properties would be absent if such activity would not have been implemented.*
2. *Unless proven otherwise, applications of structures, systems or devices with a dimension below 300 nanometres are presumed to possess such properties.*

The addition under (2) makes the delineation *functional*. With this we mean that any application within the size range is considered novel, putting the burden of proof on the shoulder of s\he who introduces or uses such applications. The delineation under (1) is not critical from a

---

[28]ISO/TC 229, Business Plan – Nanotechnologies, at p. 3.
[29]While it is interesting to follow the development of concepts under ISO we believe that we should not recommend to the international community to collectively adopt standards that are not in the public domain but available only on a commercial basis.
[30]Grobe, et al., 2008, 7.

legal viewpoint, given the addition under (2). We propose the elements (1) and (2) as a starting point for discussion and (eventually) integration in a wider context of legal structuration. As argued and included, 'size or scale' should not be bound to a definite range but manoeuvring space should be provided, as is included in the ISO definition of nano-scale[31] by using the adverb 'approximately'. Our definition is a functional translation of the operationalisation of the nanotechnology concept by Schmid et al. (2003)[32]: 'Nanotechnology is dealing with functional systems based on the use of subunits with specific size-dependent properties of the individual sub-units or of a system of those'.

However, this definition, although excellently derived from other definitions, lacks instant recognition and practical utility within the field of food and agriculture. Our proposed definition includes the four main discerned elements, and is recognisable and applicable within the context of food law and regulation.

As to Schmid et al. (referring to the given sources, pages 13–18):

Definition 1: Nanotechnology is made up of 'areas of technology where dimensions and tolerances in the range of 0.1 nm to 100 nm play a critical role'. (ref.: Glossary of the Nanoforum, www.nanoforum.org).

Definition 2: Nanotechnology, production and processing of structures and particles in the nanometer area. Nanotechnology allows for manipulation of matter on the atomic measure. The purpose is the precise structurization for the production of extremely small devices or structures with prescribed properties (ref.: Brockhaus Encyclopaedia, www.brockhaus.de, 07052003).

Definition 3: Nanotechnology describes the creation and utilization of functional materials, devices and systems with novel functions and properties that are based either on geometrical size or on material-specific peculiarities of nanostructures (ref.: www.nanoforum.org; what is Nano?).

Definition 4: Research and technology development at the atomic, molecular or macromolecular levels, in the length scale of approximately 1–100 nanometer range, to provide a fundamental understanding of phenomena and materials at the nanoscale and to create and use structures, devices and systems that have novel properties and functions because of

---

[31]Nanoscale = Size range from approximately 1–100 nm (ISO/TS 27687, 2008), but functionally be interpreted depending on the identified risks of food intake and contact with food.
[32]Schmid et al., 2003, 24.

their small and/or intermediate size (National Nanotechnology Initiative, http://www.nano.gov/omb_nifty50.htm).

Definition 5: Object of Nanotechnology is the production and application of structures, molecular materials, internal and external surfaces in critical dimensions or production tolerances of some 10 nm to atomic scales. [...] Aim is the preparation of material dependent properties of solids and their dimensions and new functionalities based on new physical-chemical-biological impact principles, caused by the submicroscopic respectively the atomic or molecular area. [...] Nanotechnology is dealing with systems with new functions and properties which depend solely on nanoscale effects of their components (ref.: Bachmann, 1998).

Definition 6: Object of Nanotechnology is the production, analysis and application of functional structures whose scales are in the area of below 100 nm. [...] An atom or a molecule does not show the physical properties we are 'used to' like electrical conductivity, magnetism, color, mechanical rigidity or a certain melting point. [...] Nanotechnology is taking place in the intermediate area between individual atoms or molecules and larger ensembles of atoms and molecules. In this area new phenomena appear which cannot be detected on macroscopic objects. (ref.: Bachmann, 2002).

Meanwhile a formal legally binding definition has been put in place in EU legislation. How does the functional definition we propose compare to this definition?

The EU definition is included in the new regulation (EU) 1169/2011 on food information, which will come into force on the 14 December 2014. Our functional definition focuses on nanotechnology. The information provisions in Regulation (EU) 1169/2011 have been drawn up to include a definition of engineered nanomaterials. Engineered nanomaterial is the result of an activity. Our functional definition focuses on the activity itself. As the functional definition is designed in terms of results (i.e. 'novel' materials), a comparison still is possible.

Article 2 (2) (t) of Regulation (EU) 1169/2011:

'engineered nanomaterial' means any intentionally produced material that has one or more dimensions of the order of 100 nm or less or that is composed of discrete functional parts, either internally or at the surface, many of which have one or more dimensions of the order of 100 nm or less, including structures, agglomerates or aggregates, which may have a size above the order of 100 nm but retain properties that are characteristic of the nanoscale.

Properties that are characteristic of the nanoscale include:

(i) *those related to the large specific surface area of the materials considered; and/or*

(ii) *specific physico-chemical properties that are different from those of the non-nanoform of the same material;*

Our functional definition has two parts of which the first has as main components:

- reductive or integrative activity
- structures, systems or devices are created
- with properties that are new
- properties would be absent if activity would not have been performed.

These elements can be discerned to a certain extent also in the food information definition on engineered nanomaterials:

Functional definition	*Regulation (EU) 1169/2011*
Reductive or integrative activity	Absent, definition refers to results
Structures, systems or devices are created	(Intentionally) produced material
At a nano-micrometre scale	May have a size above the order of 100 nm
With properties that are new	properties that are characteristic of...
Properties would be absent if...	Intentionally (produced material)

The EU definition is very open as it uses terms which have no clearly delineated content, like 'which may', 'include' or 'are characteristic'. It would therefore need to be part of a responsive environment to come to fruition. Responsiveness, however, is not immediately apparent in the structure of the EU regulation.

We have added a second part to our functional definition of nanotechnology, which does not have an equivalent in the EU definition: 'Unless proven otherwise, applications of structures, systems or devices with a dimension below 300 nanometres are presumed to possess such properties'. This implies that the initial responsibilities for applications of nanotechnology rest with the actor that makes use of them. As accountability is assigned, risk assessment and management are assigned also to the one that harvests the benefits, and that is predominantly the sponsor of the nanofood. At present, the EU definition 'engineered nanomaterials' plays exclusively a role in information law (indication 'nano' to an ingredient in the ingredient list).

As is explained in Section 5.1.2 for the purpose of labelling we mainly rely on the second part of the definition. We believe that for this purpose simplicity is more important than accuracy. For the purpose of risk assessment, however, simplicity is not the primary concern but the properties of the product. In another new EU regulation – Regulation (EU) 609/2013 on special categories of foods – the same definition from the food information regulation is also applied for the purpose for which in our opinion is more suitable: the assessment of risk (see Appendix D).

CONCLUSION

Definitions of nanotechnology use as dimensions size and/or scale, outcome, novelty of properties, causality (expressing the difference between 'natural' and engineered ('manipulated'). To use a functional delineation of nanotechnology it is not necessary to provide a 'solid' definition which holds in all applications. Instead, a delineation of the concept should be provided which puts the burden of proof on the shoulders of the actors which use manipulations of material in small dimensions with novel properties. In the practical use of the concept, details can further be enclosed to ultimately encapsulate all manifestations of applications at a small (atomic/molecular) scale with novel properties.

Lessons Learned from Selected Regulatory Reviews

C.1 INTRODUCTION

This appendix provides a summary of selected regulatory reviews of concerns over the market entry of foods to which nanotechnology has been applied, use of nanotechnology in food production, storage and tracing foods; and nanoparticles introduced into the food and environmental chain. The reviews assisted in identifying potential issues and challenges associated with the effective regulation of nanotechnology applications within the agri-food sector. Regulatory frameworks consist of principles and policies such as pre-market safety assessment, burden of proof on safety, safety evaluation and determination. Statutory frameworks established by government set out the powers of regulatory agencies, agency scope, objectives and operating boundaries.

Concerns about regulatory 'gaps', lack of confidence in regulations and regulatory agencies and need of scientific information have all been raised in the reviewed papers. Reviewed articles raised issues such as pre-market safety testing and review, impact of products and processes on the environment, the line between pharmaceuticals and nutriceuticals, animal and pesticide application and impacts. Implementation activities can include matters such as testing, measurements in the field, assessment and enforcement.

The key findings of selected regulatory reviews and a table summarising the scope or terms of reference for each review, its focus and any limitations on the work are as follows.

C.2 SELECTED REVIEWS OF REGULATORY ACTIONS ON NANOTECHNOLOGY AND FOOD

C.2.1 Regulatory Reviews Undertaken Within the EU
C.2.1.1 Chaudhry et al. (2006)[1]

In the context of a survey of the wider nanotechnology market and application, this review contains a brief section on 'food processing'[2] and 'plastics' including FCMs.[3] The Chaudhry report describes the UK and EU regulations relevant to these sectors,[4] focussing on chemical and environmental regulation.

Chaudhry concludes that there are regulatory gaps that derive either from exemptions (on a tonnage basis) under legislative frameworks or the lack of information or uncertainties over:

- a clear definition(s) encompassing the novel (or distinct) properties of nanotechnology and ENMs (i.e. whether a ENM should be considered a new or an existing material);
- current scientific knowledge and understanding of hazards and risks arising from exposure to ENMs;
- agreed dose units that can be used in hazard and exposure assessments;
- reliable and validated methods for measurement and characterisation that can be used in monitoring potential exposure to ENMs;
- potential impacts of ENMs on human and environmental health.

Chaudhry et al. conclude that:

A substantial body of work will be required to reduce these uncertainties (. . .). There is, however, an urgent need for setting clear, authoritative definitions for nanotechnologies and nanomaterials, and achieving a scientific consensus to categorise different types of nanomaterials into new (or different form) or existing substances, as this will have a major bearing on the appropriateness and applicability of current and future legislation.[5]

[1]Chaudhry et al., 2006, 22.
[2]Chaudhry et al., 2006, 22.
[3]Chaudhry et al., 2006, 25–26.
[4]Chaudhry et al., 2006, 48–51.
[5]Chaudhry et al., 2006, 8.

C.2.1.2 Chaudhry et al. (2008)

Chaudhry's second review[6] surveys current and projected processes, products and applications of nanotechnology in the food sector. It then considers the potential implications of such developments for consumer safety, in particular noting the knowledge gaps that exist in relation to safety assessment. The article examines whether existing EU food legislation is adequate to control any such risks. Manufacturers' general obligation under current law to ensure food safety, labelling and the institutional capacity of the regulator is discussed.

After reviewing the application of particular regulations to foods containing ENMs, Chaudhry et al. conclude that:

> *the gap is not necessarily a regulatory one, but potentially one of compliance by manufacturers if they have not carried out an adequate risk assessment based on data for migration, toxicity and intake. Given this, it may not be necessary to create additional/separate regulations for nanosubstances.[7]*

The ultimate responsibility for ensuring that the final food is safe rests with the seller who offers the food or packaged food for sale, or who offers unfilled materials or articles for sale to consumers for home use.[8]

In relation to labelling, Chaudhry recommends that industry takes a pro-active stance to inform, engage and consult consumers at the outset to help build public confidence, trust and acceptance. Thorough consideration and consultation with stakeholders is required. The review recommends that the food industry consider voluntarily declaring the use of nanoadditives, especially where free ENPs have been introduced into food/drinks and where such products are likely to be consumed in large quantities and/or by a large proportion of the population.

The review also concludes that:

> *[f]or a regulatory framework to be effective in controlling the potential risks from application of nanotechnology, the relevant legislation needs to provide a clear definition that encompasses the distinctive properties of nano-ingredients and additives, a clearly defined responsibility/liability for relevant*

---

[6]Chaudhry et al., 2008.
[7]Chaudhry et al., 2008, 254.
[8]Chaudhry et al., 2008, 254.

products and applications and appropriate permissible limits that relate to the (potential) effects of nano-substances in food.

C.2.1.3 Food Safety Authority of Ireland (2008)

The report by the Food Safety Authority of Ireland (FSAI)[9] provides an examination of current research and potential use of nanotechnology in food innovations. The FSAI discussed gaps in scientific knowledge and methodology regarding assessment of hazard and exposure for the purposes of risk assessment for such innovations. The use of nanotechnology in production of feed for food-producing animals and potential pathways of nanoparticles from the environment into food is reviewed. The need for better characterisation techniques for ENMs, given their use in complex environments such as biological matrices, is analysed. Regulation of foods containing ENMs is described via an outline of food laws in Ireland and the EU, with reference to pending EU regulations.

FSAI's report includes recommendations for future research. The review suggests that nanotechnology has the potential for a major impact on food innovation, particularly in intelligent FCMs and nano-encapsulated food ingredients/nutrients and sensor development. Until significant gaps in scientific knowledge and methodology regarding hazard and exposure assessment are addressed, each application of nanotechnology in food production and the wider implications of nanotechnology for the food chain should be assessed on a product-specific basis. This will also require international agreement on approaches to risk assessment to be achieved. It suggests as well that risk management approaches need to be precautionary. In this regard the FSAI concluded:

- legal provisions should be considered at EU level to ensure that food and feed produced via the application of nanotechnology are specifically controlled to ensure that where the properties are changed/re-engineered to the nanoscale, they should be re-evaluated in terms of safety;[10]
- food surveillance programmes should include investigation of potential for nanoparticles, particularly inorganic molecules such as titanium dioxide and clay particles used in packaging, to migrate into foods;

[9]Food Safety Authority of Ireland, 2008.
[10]Food Safety Authority of Ireland, 2008, 59.

- nanoparticles used in packaging could be recycled in the environment and enter the food chain indirectly;[11]
- food or food packaging in contact with food which incorporates nanoparticles should be labelled to protect and inform consumers;
- [t]here is a need for a clear and widely accepted definition of nanotechnology that differentiates between the use of nanoscale technologies in food and feed production and the actual inclusion of nanoparticles in food;[12]
- a publicly available inventory of nanotechnology-based food products/FCM be promoted by the regulator at the Irish and EU level;
- urgent consideration of whether additional controls are required in disposal and/or recycling of nanoparticle-containing food contact and other materials is further recommended.

The FSAI addressed the question of whose duty it is to determine risk. The report asks what assessment of the risk of nanotechnology-based food and feed should primarily be the responsibility of the food business operator or whether there should be a requirement for a regulatory evaluation and approval system. Many potential applications of nanotechnology in the food, feed and other industries could have the potential to swamp current regulatory approval systems. The regulator should keep industry informed of the legal requirements governing nanotechnology applications in food. The food business operators should conduct risk assessments on all such foods in order to meet their legal obligations to produce safe food. The report considers:

> that it would be a failure of "duty of care" for food companies intending to introduce nano-sized food ingredients and/or to change the natural balance of particulate size to achieve a new functionality, not to conduct an assessment of potential risk of the technology, given that it is a legal requirement for food business operators to only place safe food on the market and that safety must be demonstrable. At present, the lack of data regarding the safety of some nanoparticles in food would imply that a full risk assessment should be conducted.[13]

Finally, the FSAI concludes that food ingredients produced by nanotechnology that are covered by specific approval processes at the EU level, such as those for food additives and FCMs, should require re-evaluation of such ingredients even where there is existing

[11]Food Safety Authority of Ireland, 2008, 4.
[12]Food Safety Authority of Ireland, 2008, 53.
[13]Food Safety Authority of Ireland, 2008, 56.

authorisation for the non-nanoform of the ingredient if the nanoform is likely to alter the behaviour of the ingredient in the human body.[14] The production of a novel food by nanotechnology should require evaluation by specific approval processes for its possible risk to health before being placed on the market.[15] The FSAI questioned what percentage of a food (for instance, 5%, 10% or 50%?) should be at nanolevel for it to be defined as having been produced by nanotechnology.[16]

C.2.1.4 Food Standards Agency, UK (2008)

The link between risk assessment and regulation in the food sector is emphasised in a review of nanotechnology by the UK's Food Standards Agency (FSA).[17] The reviewed regulations require formalised 'prior approval' or placement on a 'positive list' of permitted food products or processes as well as those relying on case-by-case assessment. The FSA considers issues of openness and transparency of regulation and risk assessment. Nanofoods and processes may provide the food sector with high technical innovation. Technical innovation in nanofood and processing may give rise to competing demands for commercial confidentiality. Labelling, hygiene, animal feed and institutional capacity (in the context of ability to introduce new regulations) were reviewed by the FSA.

The review concludes that:

> *most potential uses of nanotechnologies that could affect the food area would come under some form of approval process before being permitted for use.*[18]

No major gaps in regulations are identified by the review. The reviewers noted that there is uncertainty in some areas as to whether applications of nanotechnology would be picked up consistently. The existing model for risk assessment was seen as being applicable to nanomaterials although major gaps in information for hazard identification were found. The review concluded that it would be the responsibility of manufacturers to provide the completed data.

---

[14]Food Safety Authority of Ireland, 2008, 56.
[15]Food Safety Authority of Ireland, 2008, 56.
[16]Food Safety Authority of Ireland, 2008, 56.
[17]Food Standards Agency, 2008.
[18]FSA, 2008, 2.

C.2.1.5 European Commission Communication and Staff Working Document (2008)[19]

The European Commission's Communication Regulatory aspects of nanomaterials[20] and the accompanying Commission Staff Working Document[21] 'provides a description of elements of selected EU legislation that seem most relevant and likely to apply to nanotechnologies and nano-materials'.[22] The intent of the staff working document was not to provide an in-depth review of all relevant legislative instruments across sectors. Rather, it was designed to highlight key principles and features of current EU legislative regimes relevant to human and environmental safety. The review examines the regulatory arrangements and their adequacy across a number of sectors and product applications, including industrial chemicals, worker protection, distinct classes of products and the environment. The review examined safety for foods and feed. Consideration of research needs 'to support legislative work in the field of environment, health and safety'[23] and the ways in which some of these research gaps are being addressed by the European Community and other bodies were outlined.

The EU legislative framework does not contain any regulatory provisions that specifically dealt with ENMs. Rather, the General Food Law (Regulation 178/2002) places an obligation on manufacturers to ensure the safety of their product prior to its placement on the EU Market. The General Food Law is supplemented by specific product legislation, including legislative instruments that deal specifically with novel foods,[24] FCMs,[25] food additives,[26] food supplements[27] and feed.[28]

---

[19]European Commission, 2008a.
[20]European Commission, 2008a.
[21]European Commission, 2008b.
[22]European Commission, 2008b, 4.
[23]European Commission, 2008b, 37.
[24]Regulation (EC) 258/97 concerning Novel Foods and Novel Food Ingredients (the 'Novel Foods Regulation').
[25]Framework Regulation (EC) No 1935/2004 on materials and articles intended to come into contact with food.
[26]Framework Directive 89/107/EEC (the 'Food Additives Framework Directive'). This is further supplemented by subordinate legislation. Meanwhile, the directive has been replaced by Regulation 1333/2008 on food additives.
[27]Directive 2002/46/EC on the approximation of the laws of the Member States relating to food supplements. This is further supplemented by a number of pieces of legislation.
[28]Directive 96/25/EC on the circulation of feed materials and Directive 79/373/EEC on the marketing of compound feeding stuffs. These are further supplemented by a number of other pieces of legislation.

The EU Commission concludes that the current legislative framework provides the necessary provisions for assessing possible risks associated with, for example, the use of nanomaterials which fall within the scope of the Novel Foods Regulation. Current food law provides the necessary legislative basis for the Commission 'to take, if need be, further risk management measures' in relation to labelling of such foods, 'by setting conditions of use or by information about the use of the food'.[29] The Commission further states, for example, that the requirements set down by instruments such as Regulation (EC) 1935/2004 in relation to FCMs and Directive 89/107/EEC in relation to food additives provide the 'requirements and mechanisms'[30] or 'necessary provisions'[31] to enable potential risks associated with nanomaterials in food products to be dealt with in an appropriate way. The general requirements set down by Directive 96/25 and Directive 79/373 require feed materials and feedstuff to be of 'sound, genuine and of merchantable quality'.[32] The Commission believes that the current framework provides the necessary controls to ensure the safety of feed.[33]

Overall, the Commission concluded that:

> ...current legislation covers to a large extent risks in relation to nanomaterials and that risks can be dealt with under current legislative framework. However, current legislation may have to be modified in the light of new information becoming available...[34]

This may include implementing legislation to enable authorities such as the European Food Safety Authority (EFSA) to verify the presence of nanomaterials as part of the pre-market authorisation procedures (for such products as novel foods). It may also encompass enhanced compliance procedures in relation to market surveillance activities for products which are not subject to pre-market

---

[29]European Commission, 2008b, 22.

[30]European Commission, 2008b, 23.

[31]European Commission, 2008b, 24.

[32]European Commission, 2008b, 26.

[33]While not discussed in any great detail within the context of the Commission's Communication, it is important to note that the legislative framework for foods in the EU is exceedingly dynamic, and that a number of the instruments examined by the Commission in its review have or are due to be recast. These include, for example, the Food Additives Framework Directive, which has been replaced by a common authorisation system for food additives, enzymes and flavourings, and the Novel Foods Regulation.

[34]European Commission, 2008b, 3.

authorisation processes, including food and feed generally. The Commission noted that improvements to the current state of knowledge and the implementation to legislation, including risk assessment, are pivotal to safeguarding human and environmental safety.

C.2.2 Regulatory Reviews Undertaken Within the USA
C.2.2.1 Taylor (2006)

Taylor reviewed the adequacy of the US FDA's 'legal tool kit' and other capacities for regulating the products of nanotechnology in all areas of the FDA's broad jurisdiction, including food applications.[35] Taylor analyses the laws and regulations relevant to whole foods, food ingredients, dietary supplements and food packaging materials that may have nanotechnology involvement. He also addresses manufacturers' responsibilities and the FDA's financial and scientific capacity to effectively regulate the products of nanotechnology both prior to and after their entry into the market.

Taylor's review finds significant gaps in the FDA's regulatory tools and institutional capacity for overseeing nanotechnology products. He concludes that the most fundamental step for closing the existing regulatory gaps is 'setting the criteria for determining when a nanoscale material is 'new for legal and regulatory purposes' and 'new for safety evaluation purposes".[36] Taylor's review recommends that the regulatory system should be able to '[p]lace the initial and continuing burden to demonstrate safety on the nanotechnology product's sponsor' and review the nanotechnology product's safety prior to marketing.[37] The review emphasises the need for the US Congress to provide the FDA with the resources it needs to conduct the research and build the FDA scientific capacity to understand and properly address nanotechnology safety issues.[38]

In contrast to the UK FSA review, Taylor's review recommends that the regulator has 'administrative authority to call for the submission of specified information on emerging technologies and products under its jurisdiction, including products in the development pipeline', provided such requests are focused and targeted.[39] The review notes

[35]Taylor, 2006.
[36]Taylor, 2006, 4, for example for purposes of distinguishing it from versions that are already listed in the FDA's GRAS, food additive and food packaging regulations 8.
[37]Taylor, 2006, 6.
[38]Taylor, 2006, 10.
[39]Taylor, 2006, 9.

that 'Confidentiality of trade secrets and other proprietary information must be carefully protected'.[40] Other recommendations include the need for interim pre-market notification mechanisms to address new technologies, access to companies' records concerning safety substantiation and other safety information, authority to require post-market monitoring and surveillance to check long-term safety and mandatory adverse event reporting systems appropriate for product type.[41]

C.2.2.2 US Food and Drug Administration (2007)
In 2007, the FDA Nanotechnology Task Force undertook its own in-house review of the scientific issues and key regulatory and policy challenges facing the Agency in relation to Nanotechnology. The objective of the report was to outline ways in which the FDA can both (i) enhance its knowledge of nanotechnology to support its oversight for products using such technology and (ii) inform interested stakeholders of what information may need to be developed to support the marketing of FDA-regulated products that use nanoscale materials.[42]

The report focuses on current scientific concerns, including the interaction of nanoscale materials with biological systems and the adequacy of conventional risk assessment paradigms for assessing toxicity of nanoscale materials. Many of the issues raised within this context did however relate to the 'adequacy and application'[43] of the FDA's regulatory authorities in relation to FDA-regulated products including foods, drugs, medical devices, cosmetics and veterinary products (including feed) as set out in the Federal Food, Drug and Cosmetic Act (FFDCA) and the Public Health Services Act (PHS Act). Particular attention within the regulatory policy context was given to the operation and effectiveness of pre- and post-market authorisation procedures for different classes of products, as well as issues associated with labelling of FDA-regulated products.

The report highlights the Agency's general legislative authorities, principles and rules in relation to the various product categories. It also provides, in some instances, detailed information on key sections within the FFDCA in relation to the operation of the legislative

---

[40]Taylor, 2006, 57.
[41]Taylor, 2006, 57–58.
[42]Food and Drug Administration, 2007, 4.
[43]FDA, 2007, ii.

instrument. Within this context, the FDA notes that pursuant to requirements set down in the FFDCA and subordinate instruments that:

> in all cases, whether subject to premarket authorization or not, FDA-regulated products cannot be marketed unless they satisfy specified statutory requirements. In addition to other such requirements (....) foods (including dietary requirements and food additives), color additives, and cosmetics must be safe.[44]

Looking specifically at the regulatory and policy challenges associated with the use of nanomaterials in food and feed, the FDA Task Force states that, in their opinion, the regulatory requirements set down for products requiring pre-market authorisation approval, including food, colour additives and animal feed containing a new animal drug, 'the agency's authorities [were] generally comprehensive'.[45] Moreover, the Task Force suggests that the effectiveness of the regulatory pathway was sufficient for products either manufactured using nanotechnology or incorporating nanoscale ingredients. However, it is acknowledged that additional knowledge relating to physico-chemical characteristics would be beneficial for data and risk assessment processes.

For products not subject to pre-marketing authorisation processes, including whole foods, dietary supplements and GRAS food ingredients, it is noted that the Agency's 'oversight capacity [was] less comprehensive'.[46] This was the case because the Agency does not receive, for example, safety data prior to the product's entry onto the US market. Yet as noted by the FDA Task Force, even when products are not subject to pre-market approval processes, 'manufacturers are still responsible for ensuring that the products they market are safe'.[47] This general safety requirement would apply equally to products incorporating or not incorporating nanoscale components. In recognition, however, of the limited baseline data held by the agency in relation to these products, and combined with the current scientific uncertainties associated with nanotechnology, the FDA Task Force made further recommendations. It was the recommendation that the agency work with manufacturers to increase their understanding of the effects of nanoscale materials in these products in relation to safety, and issue

[44]FDA, 2007, 20.
[45]FDA, 2007, iii.
[46]FDA, 2007, iii.
[47]FDA, 2007, 33.

new or revised guidance material in relation to additional or specific information being sought by the FDA on the use of nanoscale ingredients in GRAS food ingredients and dietary ingredients.

In considering the issue of permissible and mandatory labelling for FDA-regulated products containing nanoscale materials, including food products, the Task Force notes that there had been a request to require labelling in order to disclose ENMs within products. Pursuant to the provisions of the FFDCA, labelling of products must be truthful and contain material information, which in the case of foods, includes nutritional information or functional properties. While acknowledging that the current framework could be amended so as to require the disclosure of nanoscale materials on a label where the substance 'was a material fact for a category of product',[48] any such approach would need to be done on a case-by-case basis. Such a conclusion was based on the Task Force's view that at present the scientific data does not suggest that products containing nanoscale materials are less safe than conventional products, and as such, they did 'not believe there [was] a basis for saying that, as a general matter, a product containing nanoscale materials must be labelled as such'.[49]

Pursuant to the objectives and scope of the report, the Task Force acknowledges that while nanotechnology has the potential to impact on every class of product within its regulatory scope, in their opinion, 'materials will present regulatory challenges that are similar to those posed by other new technologies FDA has dealt with in the past'.[50] While these challenges may be amplified by nanotechnologies, especially in view of the current scientific uncertainties, they were of the view that the current regulatory framework was flexible enough to adequately regulate relevant products containing ENMs. Recommendations in the report dealt with products not subject to pre-market authorisation and ways to assist the FDA and manufacturers with ensuring the safety of such products.

C.2.2.3 Taylor (2008) Nanotech Packaging

In 'anticipation of nanotech food-packaging materials entering the regulatory process',[51] the Project on Emerging Technologies and the US

---

[48]FDA, 2007, 35.
[49]FDA, 2007, 35.
[50]FDA, 2007, 20.
[51]Project on Emerging Nanotechnologies. Taylor, 2008, 13.

Grocery Manufacturers Association initiated a joint project to consider the legal, policy, technical and scientific issues associated with the application of the FDA's regulatory process to such materials.[52] The primary purpose was 'to identify [engineered nano-scale material]-specific issues that need to be addressed to ensure that the current US regulatory system works as intended to ensure safety'.[53] A secondary objective of the study was to clarify key scientific and regulatory issues facing companies developing and commercialising food packaging materials containing ENMs. As explained by Taylor:

> the study describes and analyzes the regulatory process as applied to ENMs to a degree sufficient for its issue-identification and education purposes, but it does not make scientific or policy recommendations. The hope is that the study can help inform necessary decisions, but it does not recommend what those decisions should be.[54]

In contrast to other reviews considered within this appendix, the approach adopted in this particular study involved the expert working groups developing three distinct scenarios, each of which involved a hypothetical product which was subsequently 'run' through the regulatory system. The hypothetical products were as follows: 'active packing' that prevents contamination of the packaging itself; 'smart packaging' that detects harmful bacteria in packaged food; and improved barrier packaging for carbonated beverages.[55] Since the FDA apparently had cleared only one such product at the time of the study,[56] this approach enabled the project team to consider a range of regulatory issues that may be faced by the FDA and industry if and when such products are developed. Two important caveats of this study were that the hypothetical products 'may or may not prove technically or commercially feasible' and that the study did not provide a full "case-study" analysis of each scenario'.[57]

The report provides an overview and key aspects of the FDA's legislative competencies in relation to food packaging, under which most substances used in the packaging products are regulated as food

---

[52]Taylor, 2008.
[53]Taylor, 2008, 15.
[54]Taylor, 2008, 16.
[55]For a detailed description of each of the three hypothetical products developed for the purposes of the report, see Appendix B of Taylor (2008), 57–59.
[56]Taylor, 2008, 14.
[57]Taylor, 2008, 17.

contact substances pursuant to the FFDCA's food additive provisions, as well as those of the Environmental Protection Authority (EPA) (when anti-microbial agents and pesticide residuals are involved). Taylor notes that the overarching principle adopted by the regulatory agencies in relation to FCMs is that:

> the burden rests on the developer or other sponsor of a new packing material to demonstrate its safety, with FDA or EPA having the opportunity to review the sponsor's data prior to marketing.[58]

In accordance with the FDA's pre-authorisation procedures, manufacturers/sponsors have a legal obligation to ensure the human safety of their product prior to its placement on the US market. As Taylor illustrates, this obligation exists in relation to FCMs regardless of whether or not the FCM contains ENPs.

In examining how this pathway will operate specifically in relation to packaging products containing ENPs, a key issue identified by Taylor is whether or not nanoscale versions of previously cleared components of, for example, a food additive will be able to be marketed under an existing regulation or whether a new petition will be required. On this issue Taylor suggests that:

> a fair reading of FDA's regulations is that if there is a change in the chemical identity or composition of a listed food additive that goes significantly behind the variation covered by the petition that gave rise to the regulation, a new petition is required, even if the changed material appears to remain within the terms of the existing regulation.[59]

However, as Taylor acknowledges in the report, this argument is not in itself conclusive, and other interpretations of the regulation are possible. This was seen as significant because the

> FDA has not specifically addressed how the principles, rules and guidance outlined here would apply to nano-scale versions of previously cleared substances.[60]

The report concludes that the novel physico-chemical properties associated with ENPs will give rise to a number of issues, with the most challenging of these relating to 'how the scientific and technical

---

[58]Taylor, 2008, 19.
[59]Taylor, 2008, 34.
[60]Taylor, 2008, 35.

criteria for evaluating the food safety aspects of ENMs in food packaging will apply, in light of their novel properties'.[61] The report notes that a case-by-case approach to safety evaluation will likely remain the norm, with guidance from FDA on safety evaluation in general and how it intends to treat new nanoscale versions of previously cleared bulk substances likely to be useful.

The approach adopted by Taylor within the context of this review has a number of benefits compared to the approaches adopted by other authors. For example, by 'testing' the regulatory framework through the use of hypothetical products, Taylor was able to overcome the current limitations associated with the small number of products available in the market. Moreover, by running several different applications through the regime, Taylor highlights how the strengths and weaknesses of the framework differ in relation to the different products it must regulate.

C.2.3 Regulatory Review Undertaken Within Australia
C.2.3.1 Ludlow et al. (2007)
Ludlow et al.[62] assess how Australian regulatory frameworks can be expected to apply to nanotechnology and food products. The appropriateness of that application was analysed through five criteria: trigger and scope, requirement for regulatory approval, human safety assessment, environmental safety assessment and post-market monitoring. The review also considers manufacturers' responsibilities and labelling requirements. The Australian food regulatory framework is assessed in the review. The review provides a description of a small group of nanofood-related products.

Ludlow et al. conclude that there are six regulatory triggers common across the regulatory frameworks of consumer products including food that may fail to operate when applied to nanoproducts. Specifically in relation to food, the review also concludes that while rigorous assessment protocols for the evaluation of risks to human health are used by the regulator, current risk assessment methodologies may be inadequate for determining potential risks of food and FCM containing ENMs. Ludlow et al. conclude that it is unclear under current regulations whether existing substances reformulated at the

---

[61]Taylor, 2008, 6.
[62]Ludlow et al., 2007; Bowman and Hodge, 2007 and Hodge et al., 2007.

nanoscale level will be considered 'new' foods or FCMs. Further, it is unclear whether current methods and techniques for measuring nanomaterials will be adequate for detecting nanomaterials in food and FCMs and whether maximum limits for some foods on the basis of their weight, including additives, contaminants, natural toxins and dietary supplements, are appropriate for nanoforms of these foods.

Labelling provisions may not currently be appropriate for foods and FCMs containing ENMs for the purposes of protecting human health and safety. The review notes that there is the potential for an increasing number of products containing ENMs to sit on regulatory interfaces such as that between food and drugs, raising issues as to the boundaries of individual regulatory frameworks. Finally, it is concluded that the incorporation into national regulation of a range of international documents which may or may not take into account the potential implications of ENMs or products containing ENMs may cause inappropriate regulation of such entities in the case of food.

C.2.4 Common Themes and Areas of Agreement from the Reviews

The regulatory reviews identified many of the same issues.

A common and positive theme across reviews has been the confidence in existing food regulatory regimes. This confidence is based on the consistent finding across reviews that while nanotechnology in food and food packaging is not specifically identified in statutory frameworks, it nevertheless falls completely within the scope of existing formal regulatory and statutory authority. The obligation placed on manufactures to ensure that foods and food packaging are safe continues. And the authority held by food safety regulators around the world with respect to regulating nanotechnology in food is most often reported to be 'comprehensive'.[63] These frameworks are therefore seen at present as being both technology neutral and sufficiently robust in their application.

With the new challenges posed by the advent of nanotechnology,[64] the reviews generally conclude that statutory frameworks generally

---

[63]Clearly, this is not universal. For example, Taylor (2006) reported significant gaps in the US FDA's regulatory tools and shortcomings in its authority to call for information. Contrasting this, the subsequent FDA review, 2007, found the Agency's authorities to be 'comprehensive' for products requiring premarket authorisation, but less so for the case of whole foods and GRAS ingredients.

[64]Chaudhry et al., 2008, 256.

have the potential to evolve as new knowledge comes to light. As the European Commission noted these frameworks have the necessary provisions for assessing risks, and they can take further risk management measures if the need arises. In other words, current legislation and formal regulations can be modified as new information becomes available.

Some degree of divergence between the reviews focuses on the likely physical operational effectiveness of regulatory regimes. Some would focus more attention on issues such as measurement, standards, testing and enforcement. Some have more doubt as to the actual future effectiveness of existing regulatory regimes.

Differences of the reviews have indicated that there may be some 'gaps' between regulation and action by agencies. This may be due to the current scientific limitations associated with risk assessment. Others have also argued for more direction from the regulators to the regulated community.

One additional crucial element to the effectiveness of all regulatory regimes for nanotechnologies in food is the issue of resources. Taylor was unique in his review of this aspect for the USA and noted a gap in the FDA's capacity for overseeing ENMs. Agencies will need supporting resources and 'capability' in terms of appropriate scientific and professional expertise. Assessing sufficiency in terms of both supporting financial resources as well as available regulatory expertise clearly should form part of national but also international regulatory reviews.

Summary of Selected Regulatory Reviews				
Review	Jurisdictions	Scope/Terms of Reference	Focus	Limitations
Chaudhry et al. (2006)	UK and EU	To 'consider the appropriateness of existing regulatory frameworks for environmental regulation',[a] especially in relation to ENPs and nanotubes.	Environmental risks; but required to 'identify measures that can be put in place to ensure adequate protection for human health and the environment'.[b]	Focus is on environmental legislation.

(Continued)

(Continued)				
Review	**Jurisdictions**	**Scope/Terms of Reference**	**Focus**	**Limitations**
Taylor (2006)	USA	Adequacy of FDA's existing 'legal tools and resources for regulating the safety of nanotechnology products'.[c]	Human health and safety.	Focus is on human health and safety considerations and not environmental health and safety considerations.
FDA (2007)	USA	Evaluate the effectiveness of the current regulatory framework and approach in respect to FDA-regulated products.	Human health and safety.	Provides a general overview of the operating scope of the FDA Act and not a detailed analysis. Focus is on human health and safety considerations.
Ludlow et al. (2007)	Australia	To determine if current and forthcoming products are covered by existing regulatory frameworks, and identify products that might not be covered by the existing frameworks.[d]	Human health and safety and environmental concerns.	Could not make recommendations as to how to address gaps. Not required to make an assessment of potential risks posed by nanotechnologies.
Chaudhry et al. (2008)	EU	'A review of current and projected nanotechnology-derived food ingredients, food additives and food contact materials . . . in relation to potential implications for consumer safety and regulatory controls'.[e]	Human health and safety.	Focus is on consumer safety and not environmental safety.
European Commission (2008)	EU	To determine if nanomaterials are covered by existing EU regulatory frameworks and identify research needs.	Human health, safety and environmental aspects.	Provides general overview of operating scope of key EU legislative instruments and not a detailed analysis.
Food Standards Agency (2008)	UK and EU	'to identify potential gaps in regulation or risk assessment relating to the use of nanotechnologies and the potential deliberate or adventitious presence of	Human health and safety.	Focus is on consumer safety and not environmental safety.

(Continued)				
Review	Jurisdictions	Scope/Terms of Reference	Focus	Limitations
		manufactured nanomaterials in food'.[f]		
Food Safety Authority of Ireland (2008)	Ireland and EU	'implications for food safety of the application of nanotechnology in food production and processing. . .'[g] '. . .identify gaps in the regulatory framework and information needed to carry out an assessment of risk of nanoparticles in the food chain. . .'[h]	Human health and safety.	Focus is on consumer safety and not environmental or occupational safety.
Taylor (2008)	USA	Report explores 'legal and policy issues, as well as scientific and technical issues, that might arise in the application of the regulatory process to [engineered nanomaterials]'[i]	Human health and safety (in relation to food packaging).	Scope of review was food packaging materials.

[a]*Chaudhry et al., 2006, 6.*
[b]*Chaudhry et al., 2006, 6.*
[c]*Taylor, 2006, 15.*
[d]*Ludlow et al., 2007.*
[e]*Chaudhry et al., 2008, 241.*
[f]*FSA, 2008, 3.*
[g]*FSA, 2008, 1.*
[h]*FSA, 2008, 7.*
[i]*Taylor, 2008, 6.*

Nanotechnology in Proposed and Upcoming Regulation

Nanotechnology is on the agenda of food safety authorities and legislators. In this appendix we present some recent developments in the USA and in the EU that provide an impression of possible future developments.

D.1 DEVELOPMENTS IN THE USA

D.1.1 General

US Federal agencies assert the right to regulate nanotechnology and nano-sourced properties under a variety of Federal laws. The Environmental Protection Agency has determined that for regulatory purposes, carbon nanotubes are new chemicals under the Toxic Substances Control Act.[1] The US FDA regulates any substances added to food 'directly or indirectly' as a food additive unless the substance is GRAS,[2] a pesticide[3] or is otherwise excluded from the definition of a

---

[1]Toxic Substances Control Act of 1976, Section 5 of the Act requires pre-manufacturing notice of 'new chemical substances'. See Environmental Protection Agency, Toxic Substances Control Act Inventory Status of Carbon Nanotubes, *Federal Register* 73, and (October 31, 2008): 64946.

[2]'GRAS' is an acronym for the phrase Generally Recognized As Safe. Under Sections 201(s) and 409 of the Federal Food, Drug, and Cosmetic Act, any substance that is intentionally added to food is a food additive that is subject to premarket review and approval by FDA, unless the substance is generally recognised, among qualified experts, as having been adequately shown to be safe under the conditions of its intended use, or unless the use of the substance is otherwise excluded from the definition of a food additive. For example, substances whose use meets the definition of a pesticide, a dietary ingredient of a dietary supplement, a colour additive, a new animal drug or a substance approved for such use prior to 6 September 1958, are excluded from the definition of food additive. Sections 201(s) and 409 were enacted in 1958 as part of the Food Additives Amendment to the Act. While it is impracticable to list all ingredients whose use is generally recognised as safe, FDA published a partial list of food ingredients whose use is generally recognised as safe to aid the industry's understanding of what does not require approval. Food, Drug & Cosmetic Act Sec. 201(s).

[3]Pesticide residues are regulated with tolerance levels established by US Environmental Protection Agency. Fortin, 2009, 212.

food additive.[4] Components of FCMs that can be expected to migrate to food, any substance added directly to food and substances that may become components of food as a part of processing process require premarket approval based on safety.[5] 'Where appropriate to ensure safety, the FDA places limitations on the physical and chemical properties of food additives, which include particle size' according to a FDA's 2007 report.[6] Likewise, the FDA that regulates any substance capable of imparting colour to any food[7] requires premarket approval based on safety. The FDA lacks authority for premarket authorisation for dietary supplements[8] based on safety. Recently enacted legislation will give the FDA greater access to adverse events for dietary supplements.[9] New animal drugs, including new animal drugs for use in animal feed, are subject to a premarket authorisation process intended to establish the products' safety.[10]

D.1.2 US FDA Nanotechnology Task Force Report 2007
D.1.2.1 Recommendations for Consideration[11]
The Task Force recommends strengthening the FDA's promotion of, and participation in, research and other efforts to increase scientific understanding, to facilitate assessment of data needs for regulated products. Such activities should, where appropriate, be coordinated with and leveraged against activities supported by other Federal agencies, the private sector or other countries. This would include:

• promoting efforts, and participating in collaborative efforts, to further understanding of biological interactions of nanoscale materials,

---

[4]Fortin, 2009, 272.

[5]Section 409 of the Federal Food Drug and Cosmetic Act and 21 CFR Parts 170 and 171 describe in general terms the information and data necessary to establish the safety of food additives and ingredients. These authorities are supplemented by technical guidance documents providing more specific data recommendations. In addition, FDA may require any other information that it determines during its review which is necessary to establish safety. See also Fortin, 2009, 272–278.

[6]US Food and Drug Administration Nanotechnology Task Force Report 2007.

[7]FFDCA Sections 201(t) and 721 cover colour additives and safety of the colour additive.

[8]US Food and Drug Administration Nanotechnology Task Force Report 2007.

[9]Beginning 22 December 2007, any serious adverse events associated with a dietary supplement reported to the product's manufacturer, packer or distributor will have to be submitted to FDA. In addition, businesses will have to keep records of all dietary supplement adverse events, serious or non-serious, and FDA will have access to those records during inspections. US Food and Drug Administration Nanotechnology Task Force Report 2007.

[10]FDCA Section 512.

[11]US Food and Drug Administration Nanotechnology Task Force Report 2007.

including, as appropriate the development of data to assess likelihood of long-term health effects from exposure to specific nanoscale materials;

- assessing data on general particle interactions with biological systems and on specific particles of concern to the FDA;
- promoting and participating in collaborative efforts, to further understanding of the science of novel properties that might contribute to toxicity, such as surface area or surface charge;
- promoting and participating in collaborative efforts to further understanding of measurement and detection methods for nanoscale materials;
- collecting/collating/interpreting scientific information, including use of data calls for specific product review categories (. . .);
- building in-house expertise;
- building infrastructure to share and leverage knowledge internally and externally, seeking to collect, synthesise and build upon information from individual studies of nanoscale materials;
- ensuring consistent transfer and application of relevant knowledge through establishment of an agency-wide regulatory science coordination function for products containing nanoscale materials.

To be marketed, FDA-regulated products must be safe and, as applicable, effective. FDA-regulated products must also meet all applicable good manufacturing practice and quality requirements. Adequate testing methods are needed regardless of whether a product is subject to premarket authorisation or not. Accordingly, the following recommendations are relevant to all categories of FDA-regulated products. The agency should:

- evaluate the adequacy of current testing approaches to assess safety, effectiveness and quality of products that use nanoscale materials;
- promote and participate in the development of characterisation methods and standards for nanoscale materials;
- promote and participate in the development of models for the behaviour of nanoscale particles *in vitro* and *in vivo*.

The Task Force recommends encouraging manufacturers to consult with the agency regarding the appropriateness of testing methodologies for evaluating products using nanoscale materials.

D.1.3 The Nanotechnology Safety Act of 2010

On Thursday 21 January 2010, Senator Mark Pryor (D-Ark) introduced S. 2942, 'The Nanotechnology Safety Act of 2010'. The bill is co-sponsored by Senator Benjamin Cardin (D-MD).

It mainly calls upon the FDA to ensure it is scientifically up to date.

As to the status of this bill, it must be remarked that it is common place for US Senators to submit bills for consideration. The majority of these are never enacted.

SECTION 1. SHORT TITLE.
This Act may be cited as the "Nanotechnology Safety Act of 2010".
SEC. 2. NANOTECHNOLOGY PROGRAM.
Chapter X of the Federal Food, Drug, and Cosmetic Act (21 U.S.C. 391 et seq.) is amended by adding at the end the following:
SEC. 1011. NANOTECHNOLOGY PROGRAM.
(a) IN GENERAL.—Not later than 180 days after the date of enactment of the Nanotechnology Safety Act of 2010, the Secretary of Health and Human Services, in consultation with the Secretary of Agriculture, shall establish within the Food and Drug Administration a program for the scientific investigation of nanoscale materials included or intended for inclusion in FDA-regulated products, to address the potential toxicology of such materials, the effects of such materials on biological systems, and interaction of such materials with biological systems.
(b) PROGRAM PURPOSES.—The purposes of the program established under subsection (a) shall be to—
(1) assess scientific literature and data on general nanoscale material interactions with biological systems and on specific nanoscale materials of concern to Food and Drug Administration;
(2) develop and organize information using databases and models that will enable the behavior of classes of nanoscale materials with biological systems;
(3) promote intramural Administration programs and participate in collaborative efforts, to further the understanding of the science of novel properties at the nanoscale that might contribute to toxicity;
(4) promote and participate in collaborative efforts to further the understanding of measurement and detection methods for nanoscale materials;
(5) collect, synthesize, interpret, and disseminate scientific information and data related to the interactions of nanoscale materials with biological systems;
(6) build scientific expertise on nanoscale materials within such Administration;
(7) ensure ongoing training, as well as dissemination of new information within the centers of such Administration, and more broadly across such

Administration, to ensure timely, informed consideration of the most current science;

(8) encourage such Administration to participate in international and national consensus standards activities; and

(9) carry out other activities that the Secretary determines are necessary and consistent with the purposes described in paragraphs (1) through (8).

(c) PROGRAM ADMINISTRATION.—

(1) PROGRAM MANAGER.—In carrying out the program under this section, the Secretary shall designate a program manager who shall supervise the planning, management, and coordination of the program.

(2) DUTIES.—The program manager shall—

(A) develop a detailed strategic plan for achieving specific short- and long-term technical goals for the program;

(B) coordinate and integrate the strategic plan with investments by the Food and Drug Administration and other departments and agencies partici-pating in the National Nanotechnology Initiative; and

(C) develop intramural Administration programs, contracts, memoranda of agreement, joint funding agreements, and other cooperative arrangements necessary for meeting the long-term challenges and achieving the specific technical goals of the program.

(d) REPORTS.—Not later than March 1, 2012 and March 1, 2014, the Secretary shall submit to the Committee on Health, Education, Labor, and Pensions and the Committee on Appropriations of the Senate and the Committee on Energy and Commerce and the Committee on Appropriations of the House of Representatives a report on the program carried out under this section. Such report shall include—

(1) a review of the specific short- and long-term goals of the program;

(2) an assessment of current and proposed funding levels for the program, including an assessment of the adequacy of such funding levels to support program activities; and

(3) a review of the coordination of activities under the program with other departments and agencies participating in the National Nanotechnology Initiative.

(e) AUTHORIZATION OF APPROPRIATIONS.—

There are authorized to be appropriated to carry out this section, $25,000,000 for each of fiscal years 2011 through 2015. Amounts appropriated pursuant to this subsection shall remain available until expended.

D.2 DEVELOPMENTS IN THE EU

D.2.1 Novel Foods

In 2006 the EC initiated a review of the Novel Foods Regulation (Regulation 258/97). On 14 January 2008 the Commission submitted a proposal (COM(2007) 872) to the European Parliament and to the Council. The legislative procedure encompasses two readings. After the first reading, both the Parliament and the Council proposed amendments. Some of these aim to bring nanofoods within the ambit of the Novel Foods Regulation. Here below we include the provisions on nanotechnology proposed by the European Parliament (section II) and by the Council (section III). The proposal has been rejected by the European Parliament. It is expected that the European Commission will submit a new proposal. As far as nanofood is concerned, the new proposal is expected to be similar to the old.

D.2.2 European Parliament Legislative Resolution of 25 March 2009[12]

Article 2

Scope

1. This Regulation shall apply to the placing of novel foods on the market in the Community.

2. (...)

3. Notwithstanding paragraph 2, this Regulation shall apply to food additives, food enzymes, flavourings and certain food ingredients with flavouring properties to which a new production process not used before 15 May 1997 is applied that gives rise to significant changes in the composition or structure of the food, such as engineered nanomaterials.

Article 3

Definitions

2. The following definitions shall also apply:

(a) 'novel food' means:

(iv) food containing or consisting of engineered nanomaterials not used for food production within the Community before 15 May 1997.

(f) 'engineered nanomaterial' means any intentionally produced material that has one or more dimensions of the order of 100 nm or less or is composed of discrete functional parts, either internally or at the surface, many of which have one or more dimensions of the order of 100 nm or less, including structures, agglomerates or aggregates, which may have a size above the order of 100 nm but retain properties that are characteristic to the nanoscale.

Properties that are characteristic to the nanoscale include:

(i) those related to the large specific surface area of the materials considered and/or

(ii) specific physico-chemical properties that are different from those of the non-nanoform of the same material.

3. In view of the various definitions of nanomaterials published by different bodies at international level and the constant technical and scientific developments in the field of nanotechnologies, the Commission shall adjust and adapt point (f) of paragraph 2 to technical and scientific progress and with

---

[12]Official Journal 6.5.2010, C 117 E/236.

definitions subsequently agreed at international level. That measure, designed to amend non-essential elements of this Regulation, shall be adopted in accordance with the regulatory procedure with scrutiny referred to in Article 20(3).

Article 7

Conditions for inclusion in the Community list

2. Foods to which production processes have been applied that require specific risk assessment methods (e.g. foods produced using nanotechnologies) may not be included in the Community list until such specific methods have been approved for use, and an adequate safety assessment on the basis of those methods has shown that the use of the respective foods is safe.

Article 8

Content of the Community list

6. All ingredients present in the form of nanomaterials shall be clearly indicated in the list of ingredients. The names of such ingredients shall be followed by the word 'nano' in brackets.

D.2.3 Position (EU) No 6/2010 of the Council at First Reading[13]

Article 3

Definitions

1. (. . .)

2. The following definitions shall also apply:

(a) 'novel food' means food that was not used for human consumption to a significant degree within the Union before 15 May 1997, including:

(iv) food containing or consisting of engineered nanomaterials;

(c) 'engineered nanomaterial' means any intentionally produced material that has one or more dimensions of the order of 100 nm or less or that is composed of discrete functional parts, either internally or at the surface, many of which have one or more dimensions of the order of 100 nm or less, including structures, agglomerates or aggregates, which may have a size above the order of 100 nm but retain properties that are characteristic of the nanoscale.

Properties that are characteristic of the nanoscale include:

(i) those related to the large specific surface area of the materials considered; and/or

(ii) specific physico-chemical properties that are different from those of the non-nanoform of the same material;

D.2.4 Commission Recommendation on the Definition of Nanomaterial (2011/696/EU)

On 18 October 2011 the EC issued a recommendation on the definition of nanomaterial.

---

[13]Official Journal 11.5.2010, C 122 E/38.

1. (. . .)

2. 'Nanomaterial' means a natural, incidental or manufactured material containing particles, in an unbound state or as an aggregate or as an agglomerate and where, for 50 % or more of the particles in the number size distribution, one or more external dimensions is in the size range 1 nm−100 nm.

In specific cases and where warranted by concerns for the environment, health, safety or competitiveness the number size distribution threshold of 50 % may be replaced by a threshold between 1 and 50 %.

D.2.5 Regulation (EU) 1169/2011 on Food Information to Consumers

A new regulation on labelling and other forms of provision of food information to consumers, will enter into force on 14 December 2014.

Article 3
Definitions
1. (. . .)

2. The following definitions shall also apply:

(t) 'engineered nanomaterial' means any intentionally produced material that has one or more dimensions of the order of 100 nm or less or that is composed of discrete functional parts, either internally or at the surface, many of which have one or more dimensions of the order of 100 nm or less, including structures, agglomerates or aggregates, which may have a size above the order of 100 nm but retain properties that are characteristic of the nanoscale.

Properties that are characteristic of the nanoscale include:

(i) those related to the large specific surface area of the materials considered; and/or

(ii) specific physico-chemical properties that are different from those of the non-nanoform of the same material.

Article 18
List of ingredients
1. (. . .)

2. (. . .)

3. All ingredients present in the form of engineered nanomaterials shall be clearly indicated in the list of ingredients. The names of such ingredients shall be followed by the word 'nano' in brackets.

D.2.6 Regulation (EU) 609/2013 on Food for Special Purposes

A new regulation on PARNUTS will enter into force on 20 July 2016. It adopts the definition of ENM from Regulation 1169/2011.

Article 18
General compositional and information requirements

1. (...)

2. Food referred to in Article 1(1) shall not contain any substance in such quantity as to endanger the health of the persons for whom it is intended.

For substances which are engineered nanomaterials, compliance with the requirement referred to in the first subparagraph shall be demonstrated on the basis of adequate test methods, where appropriate.

D.3 CURRENT DEVELOPMENTS

From the above the impression can be derived that the USA mainly plans to deal with nanotechnology within the existing regulatory framework. Additional efforts are focussed on increasing understanding.

Proposals in the EU aim to ensure that food applications of nanotechnology fulfil the definition of novelty and thus fall within the scope of the premarket approval requirement for novel foods.

As from 2014 presence of ENMs in food has to be declared on the label in the EU.

APPENDIX *E*

Research Team

H.J. (Harry) Bremmers
Associate Professor of Law and Economics at Wageningen
University (the Netherlands) <www.law.wur.nl>
H. (Hans) Bouwmeester
Senior scientist nanotoxicology at RIKILT Institute of Food Safety
part of Wageningen University and Research Center (the
Netherlands) <www.rikilt.wur.nl/UK/>
L.L. (Leon) Geyer
Professor, Environmental and Agricultural Law and Economics
Department of Agricultural and Applied Economics, at Virginia
Polytechnic Institute and State University (Virginia Tech)
(Blacksburg, VA) <http://www.cals.vt.edu/departments/aaec.html>
N. (Nidhi) Gupta
PhD Researcher, Marketing and Consumer Behaviour Group at
Wageningen University (the Netherlands) <http://www.mcb.wur.
nl>
I. (Iwona) Kozak
Master student in Food Quality Management Governance at
Wageningen University (the Netherlands) <www.mfq.wur.nl>
B.M.J. (Bernd) van der Meulen
Professor of Law and Governance at Wageningen University (the
Netherlands) <www.law.wur.nl>; director of the European
Institute for Food Law <www.food-law.nl>; member of the board
of directors of the European Food Law Association <www.EFLA-
AEDA.org> and member or the board of editors of the European
Food and Feed Law Review <www.lexxion.eu/effl>

M.P. (Margherita) Poto
Postdoc Food Law at Wageningen University (the Netherlands)
<www.law.wur.nl>
K.P. (Kai) Purnhagen
Assistant Professor of Law and Governance at Wageningen University (the Netherlands) and distinguished international visitor at the Erasmus University of Rotterdam, Rotterdam Institute of Law and Economics

REFERENCES

All hyperlinks were lastly accessed between 5 and 25 September 2013.

SOURCES OF LAW

Australia, FSANZ Standard 3.2.3, available at: <http://www.foodstandards.gov.au/foodstandards/foodsafetystandardsaustraliaonly/standard323.cfm>.

Australia/New Zealand, FSANZ standard 1.5.1, available at: <http://www.foodstandards.gov.au/foodstandards/foodsafetystandardsaustraliaonly/standard151.cfm>.

EU, Court of First Instance, Case T-326/07, *Cheminova* v. *Commission*, available at: <http://eur-lex.europa.eu/en/index.htm>.

EU, Directive 79/373, available at: <http://eur-lex.europa.eu/en/index.htm>.

EU, Directive 89/107, available at: <http://eur-lex.europa.eu/en/index.htm>.

EU, Directive 96/25, available at: <http://eur-lex.europa.eu/en/index.htm>.

EU, Regulation 258/97, available at: <http://eur-lex.europa.eu/en/index.htm>.

EU, Regulation 178/2002, available at: <http://eur-lex.europa.eu/en/index.htm>.

EU, Regulation 1935/2004, available at: <http://eur-lex.europa.eu/en/index.htm>.

EU, Regulation 853/2004, available at: <http://eur-lex.europa.eu/en/index.htm>.

EU, Regulation 1333/2008, available at: <http://eur-lex.europa.eu/en/index.htm>.

UN, International Covenant on Economic, Social and Cultural Rights (ICESCR), available at <http://www2.ohchr.org/english/law/cescr.htm>.

USA, FFDCA [21 USC] available at: <http://www.gpoaccess.gov/uscode/>.

USA, Toxic Substances Control Act of 1976, available at: <http://www.gpoaccess.gov/uscode/>.

WTO, General Agreement on Tariffs and Trade (GATT), available at: <http://www.wto.org/index.htm>.

WTO, Agreement on the Application of Sanitary and Phytosanitary Measures (SPS Agreement), available at: <http://www.wto.org/index.htm>.

SOURCES OF SOFT LAW

Codex Alimentarius Commission, Food Hygiene. Basic Texts, fourth edition, WHO/FAO Rome 2009, available at: <ftp://ftp.fao.org/codex/Publications/Booklets/Hygiene/FoodHygiene_2009e.pdf>.

Codex Alimentarius Commission, General Standard for Food Additives, CODEX STAN 192-1995 (lastly revised 2009) <http://www.codexalimentarius.net/gsfaonline/CXS_192e.pdf>.

Codex Alimentarius Commission, General Standard for the Labelling of Prepackaged Foods, CODEX STAN 1-1985 (Rev. 1-1991), lastly amended in 2005, available at: <ftp://ftp.fao.org/codex/Publications/Booklets/Labelling/Labelling_2007_EN.pdf>.

Codex Alimentarius Commission, Procedural Manual, 21st edition, Rome 2013, available at: <ftp://ftp.fao.org/codex/Publications/ProcManuals/Manual_21e.pdf>.

Codex Alimentarius Commission, Recommended International Code of Practice General Principles of Food Hygiene, CAC/RCP 1-1969, Rev. 4 (2003), available at: <ftp://ftp.fao.org/codex/Publications/Booklets/Hygiene/FoodHygiene_2009e.pdf>.

FAO, Voluntary Guidelines, to support the progressive realization of the right to adequate food in the context of national food security, Adopted by the 127th Session of the FAO Council November 2004, Food and Agriculture Organization of the United Nations, Rome, 2005, available at: <http://www.fao.org/righttofood/publi_01_en.htm>.

UN, Committee on Economic, Social and Cultural Rights, General Comment 3 on the nature of State parties' obligations, available at: <http://www2.ohchr.org/english/bodies/cescr/comments.htm>.

UN, Committee on Economic, Social and Cultural Rights, General Comment 12 on the right to adequate food, available at: <http://www2.ohchr.org/english/bodies/cescr/comments.htm>.

POLICY DOCUMENTS

EU, Council, New Novel Foods Regulation, Position (EU) No 6/2010 of the Council at First Reading, Official Journal 11.5.2010, C 122 E/38 <http://eur-lex.europa.eu/LexUriServ/LexUriServ.do?uri = OJ:C:2010:122E:0038:0057:EN:PDF>.

EU, European Commission, White Paper on Food Safety (COM(1999) 719, available at: <http://ec.europa.eu/dgs/health_consumer/library/pub/pub06_en.pdf>.

EU, European Commission, Communication on the precautionary principle, COM(2000) 1, available at <http://ec.europa.eu/dgs/health_consumer/library/pub/pub07_en.pdf>.

EU, European Commission (2008a), *Regulatory Aspects of Nanomaterials*. COM(2008) 366 final, Brussels: EC <http://eur-lex.europa.eu/LexUriServ/LexUriServ.do?uri = COM:2008:0366:FIN:en:PDF>.

EU, European Commission (2008b), *Regulatory Aspects of Nanomaterials: Summary of legislation in relation to health, safety and environment aspects of nanomaterials, regulatory research needs and related measures.* SEC(2008) 2036, Brussels: EC <http://www.euractiv.com/31/images/SEC(2008)%202036_tcm31-173474.pdf>.

EU, European Commission (2009) Communication from the commission to the council, the European Parliament and the European Economic and Social Committee. Nanosciences and Nanotechnologies: An action plan for Europe 2005–2009. Second Implementation Report 2007–2009, Brussels, 29.10.2009, COM(2009) 607 final <http://ec.europa.eu/nanotechnology/pdf/comm_2008_0366_en.pdf>.

EU, European Commission, Consultation Paper on Nanotechnologies: "A European Code of Conduct for Responsible Nanosciences and Nanotechnologies Research is part of the European Commission's ambition to promote a balanced diffusion of information on nanosciences and nanotechnologies and to foster an open dialogue, involving the broadest possible range of interested parties," available at: <http://ec.europa.eu/research/consultations/pdf/nano-consultation_en.pdf>.

EU, European Food Safety Authority (EFSA), 6th Scientific Colloquium Report – Risk-benefit analysis of foods: methods and approaches, July 2007 <http://www.efsa.europa.eu/en/colloquiareports/colloquiariskbenefit.htm>.

EU, EFSA, The Potential Risks Arising from Nanoscience and Nanotechnologies on Food and Feed Safety. Scientific Opinion of the Scientific Committee, 10 February 2009, The EFSA J 958:1–39 <http://www.efsa.europa.eu/en/scdocs/scdoc/958.htm>.

EU, European Parliament, New Novel Foods Regulation, legislative resolution of 25 March 2009, Official Journal 6.5.2010, C 117 E/236 <http://eur-lex.europa.eu/LexUriServ/LexUriServ.do?uri = OJ:C:2010:117E:0236:0254:EN:PDF>.

EU, Scientific Committee on Emerging and Newly Identified Health Risk (SCENIHR). 2007. Opinion on: The appropriateness of the risk assessment methodology in accordance with the technical guidance documents for new and existing substances for assessing the risks of nanomaterials European Commission Health & Consumer Protection Directorate-General. Directorate C − Public Health and Risk Assessment C7 − Risk Assessment <http://ec.europa.eu/enterprise/newsroom/cf/document.cfm?action = display&doc_id = 3171&userservice_id = 1>.

EU, SCENIHR, 2008. Opinion on: The scientific aspects of the existing and proposed definitions relating to products of nanoscience and nanotechnologies European Commission Health and Consumer Protection Directorate-General. Directorate C − Public Health and Risk Assessment C7 − Risk Assessment <http://ec.europa.eu/health/ph_risk/committees/04_scenihr/docs/scenihr_o_012.pdf>.

FAO, Jessica Vapnek and Melvin Spreij, Perspectives and guidelines on food legislation, with a new model food law, FAO Legislative Study 87, Rome 2005, available at: <ftp://ftp.fao.org/docrep/fao/009/a0274e/a0274e00.pdf>.

FAO/WHO publication (2003). Assuring Food Safety and Quality: Guidelines for Strengthening National Food Control Systems, available at: <http://www.who.int/foodsafety/publications/fs_management/guidelines_foodcontrol/en/index.html>; <http://www.fao.org/docrep/006/y8705e/y8705e00.HTM>.

FAO/WHO Expert Meeting on the Application of Nanotechnologies in the Food and Agriculture Sectors: Potential Food Safety Implications Meeting Report 1−5 June 2009 <http://www.fao.org/ag/agn/agns/files/FAO_WHO_Nano_Expert_Meeting_Report_Final.pdf>.

Ireland, Food Safety Authority of Ireland (2008). *The Relevance for Food Safety of Applications of Nanotechnology in the Food and Feed Industries.* Dublin: FSAI.

UK, Food Standards Agency (2008). *A Review of the Potential Implications of Nanotechnologies for Regulations and Risk Assessment in Relation to Food.* London: FSA.

UK, Food Standards Agency, Report of FSA Regulatory Review, August 2008. <http://www.food.gov.uk/multimedia/pdfs/nanoregreviewreport.pdf>.

UK, Her Majesties Government, UK Nanotechnologies Strategy, Small Technologies, Great Opportunities, March 2010, <http://www.bis.gov.uk/assets/biscore/corporate/docs/n/10-825-nanotechnologies-strategy>/<http://interactive.bis.gov.uk/nano/>.

USA, Environmental Protection Agency, Toxic Substances Control Act Inventory Status of Carbon Nanotubes, *Federal Register* 73, and (October 31, 2008): 64946.

USA, Food and Drug Administration (2007), *Nanotechnology − A Report of the US Food and Drug Administration Nanotechnology Task Force.* Washington, DC, FDA <http://www.fda.gov/ScienceResearch/SpecialTopics/Nanotechnology/NanotechnologyTaskForceReport2007/default.htm#ability>.

PRIVATE STANDARDS

ISO, 2008. Guideline TS 27687.

ISO/TC 229, Business Plan − Nanotechnologies.

LITERATURE

Arcuri, A., 2000. Product safety regulation. In: Bouckaert, B., De Geest, G. (Eds.), Encyclopedia of Law and Economics, Volume III. The Regulation of Contracts. Edward Elgar, p. 329. Available at: <http://encyclo.findlaw.com/5130book.pdf>.

Bawa, R.B., Bawa, S.R., Maebius, S.B., Flynn, T., Wei, C.h., 2005. Protecting new ideas and inventions in nanomedicine with patents. Nanomed. Nanotech. Biol. Med. 1, 150−158. Available at: <http://www.foley.com/files/tbl_s31Publications/FileUpload137/2879/Protecting%20Nanomedicine%20Inventions.pdf>.

Black, J., 2001. Decentring regulation: understanding the role of regulation and self regulation in a "Post Regulatory" World. Curr. Leg. Probl. 54, 103−147.

Black, J., 2008. Forms and paradoxes of principles-based regulation. Capital Mark. Law J. 3, 425.

Black, J., Hopper, M., Band, C., 2007. Making a success of principles-based regulation. Law Financ. Mark. Rev., 191.

Boisrobert, C., Stjepanovic, A., Oh, S., Lelieveld, H. (Eds.), 2009. Ensuring Global Food Safety, Exploring Global Harmonization. Elsevier, London, UK.

Bouwmeester, H., Dekkers, S., Noordam, M.Y., Hagens, W.I., Bulder, A.S., de Heer, C., et al., 2009. Review of health safety aspects of nanotechnologies in food production. Regul. Toxicol. Pharmacol. 53, 52–62. Available at: <http://www.ncbi.nlm.nih.gov/pubmed/19027049>.

Bowman, D.M., Hodge, G.A., 2007. A small matter of regulation: an international review of nanotechnology regulation. Columbia Sci. Technol. Law Rev. 8, 1–36. Available at: <http://www.stlr.org/volumes/volume-viii-2006-2007/bowman/>.

Bowman, D.M., Hodge, G.A., 2007. A small matter of regulation: An international review of nanotechnology regulation. Columbia Sci. Technol. Law Rev. 8, 1–36.

Bowman, D.M., D'Silva, J., van Calster, G., 2010. Defining nanomaterials for the purpose of regulation within the European Union. Eur. J. Risk Regul., 115–121.

Bradford, A., 2012. The Brussels effect. Northwest. Univ. Law. Rev. 107, 1.

Bremmers, H., van der Meulen, B.M.J., Purnhagen, K.P., 2013. Multi-Stakeholder Responses to the European Health Claims Requirements – A Law and Economics Assessment of Regulation (EU) 1924/2006, Wageningen Working Paper in Law and Governance No. 2013/01, 1.

Carlson, S., Carvajal, R., Coutrelis, N., Desjeux, J.F., Morelli, L., Van Dael, P., et al., 2010a. Publish and Perish: a disturbing trend in the European Union's regulation of nutrition health claims made on foods. Update Magazine/FDLI 5, 50–52.

Carlson, S.R. Carvajal, N. Coutrelis, J.F. Desjeux, L. Morelli, L. B.M.J.van der Meulen, et al., 2010b. Protection of proprietary data. Why published data should not be excluded from protection under Article 21 of Regulation 1924/2006, Eur. Food Feed Law Rev., 3: 166-172

Chau, C.F., Wu, S.H., Yen, G.C., 2007. The development of regulations for food nanotechnology. Trends Food Sci. Technol. 18 (5), 269–280.

Chaudhry, Q., Blackburn, J., Floyd, P., George, C., Nwaogu, T., Boxall, A., et al., 2006. Final Report: A Scoping Study to Identify Gaps in Environmental Regulation for the Products and Applications of Nanotechnologies. Defra, London.

Chaudhry, Q., Scotter, M., Blackburn, J., Ross, B., Boxall, A., Castle, L., et al., 2008. Applications and implications of nanotechnologies for the food sector. Food Addit. Contam. Rev. 25 (3), 241–258.

Coffee, J.C., 1987. The future of corporate federalism: state competition and the new trend toward de facto federal minimum standards. Cardozo Law Rev. 8, 759.

D'Silva, J., 2011. What's in a name? – defining a 'nano-material' for regulatory purposes in Europe. Eur. J. Risk Regul., 85.

Drexler, E., Peterson, C., Pergamit, G., 1991. Unbounding the Future: The Nanotechnology Revolution. Foresight Nanotech Institute.

ETC Group, 2006. Action Group on Erosion, nanotech product recall underscores need for nanotech moratorium: is the magic gone? Technol. Concentration. Available at: <http://www.etcgroup.org/upload/publication/14/01/nrnanorecallfinal.pdf>.

Ford, C., 2008. New governance, compliance and principles-based securities regulation. Am. Bus. Law J. 45, 1–60.

Forrest, D., 1989. Regulating nanotechnology development. Available at: <http://research.lifeboat.com/forrest.htm>.

Fortin, N.D., 2009. Food Regulation, Law, Science, Policy, and Practice. John Wiley and Sons, Inc. Hoboken, New Jersey.

Fortin, N.D., 2011. The United States FDA Food Safety Modernization Act: the key new requirements. Eur. Food Feed Law Rev., 260–268.

Hodge, G., Bowman, D., Ludlow, K., 2007. New Global Frontiers in Regulation: The Age of Nanotechnology. Edward Elgar, Cheltenham, UK.

Hunt, W.H., 2004. Nanomaterials: nomenclature, novelty, and necessity. J. Miner. Metals Mater. Soc. 56, 13–18.

Kallet, A., Klink, F., 1933. 100 000 000 Guinea Pigs. Vanguard Press, New York.

Ludlow, K., Bowman, D.M., Hodge, G.A., 2007. A Review of Possible Impacts of Nanotechnology on Australia's Regulatory Framework, Monash Centre for Regulatory Studies, Melbourne, Australia.

Mehta, M.D., 2004. From biotechnology to nanotechnology: what can we learn from earlier technologies? Bull. Sci. Technol. Soc. 24 (1), 34–39.

Meulen, B.v.d., Velde, M.v.d., 2008. European Food Law Handbook. Wageningen Academic Publishers, Wageningen, The Netherlands.

Nel, A., Xia, T., Madler, L., Li, N., 2006. Toxic potential of materials at the nanolevel. Science 311 (5761), 622–627.

Otsuki, T., Wilson, J.S., Sewadeh, M., 2001. Saving two in a billion: quantifying the trade effect of European food safety standards on African exports. Food Policy 26, 495–514. Available at: <www.elsevier.com/locate/foodpol>.

Renn, O., 2006. Nanotechnology and the need for risk governance. Journal of Nanoparticle Research 8 (2), 153–191.

Roszek, B., de Jong, W.H., Geertsman, R.E., 2005. RIVM report 265001001: Nanotechnology in medical applications: state-of-the-art in materials and devices. RIVM, the Netherlands. Available at: <http://www.rivm.nl/bibliotheek/rapporten/265001001.pdf>.

Sanguansri, P., Augustin, M.A., 2006. Nanoscale materials development-a food industry perspective. Trends Food Sci. Technol. 17, 547–556. Available at: <http://www.sciencedirect.com/science?>_ob = ArticleURL&_udi = B6VHY-4K0D7WR-1&_user = 10&_coverDate = 10%2F31%2F2006&_rdoc=1&_fmt=high&_orig=search&_sort=d&_docanchor = &view = c&_searchStrId=1345426594&_rerunOrigin=google&_acct=C000050221&_version=1&_urlVersion=0&_userid = 10&md5 = 85569be2b402a7e4a7c8016505c78917>.

Schmid, G., Decker, M., Ernst, H., Fuchs, H., Grünwald, W., Grunwald, A., et al., 2003. Small Dimensions and Material Properties: A Definition of Nanotechnology. Europäische Akademie zur Erforschung von Folgen wissenschaftlich-technischer Entwicklungen. Bad Neuenahr-Ahrweiler GmbH. Graue reihe, nr. 35 <http://www.ea-aw.de/fileadmin/downloads/Graue_Reihe/GR_35_Nanotechnology_112003.pdf>.

Schomberg, R., Davies, S. (Eds.), 2010. Understanding Public Debate on Nanotechnologies Options for Framing Public Policy. A report from the European Commission Services. Available at: <http://www.nanotechproject.org/process/assets/files/8304/debate_nano_100203.pdf>.

Segarra, A.E., Rawson, J.M., 2001. CRS Report for Congress, StarLink Corn Controversy: Background. Available at: <http://research.policyarchive.org/3402.pdf>.

Siegrist, M., Cousin, M-E., Kastenholz, H., Wiek, A, 2007. Public acceptance of nanotechnology foods and food packaging: the influence of affect and trust. Appetite 49, 459–466. Available at: <http://www.empa.ch/plugin/template/empa/*/63901>.

Szajkowska, A., 2012. Regulating food law. Risk Analysis and the Precautionary Principle as General Principles of EU Food Law. Wagenningen Academic Publishers.

Taniguchi, N., 1974. On the basic concept of nanotechnology. Proceedings of the International Conferences on Product Engineering, Tokio, Japan.

Taylor, M.R., 2006. Regulating the Products of Nanotechnology: Does FDA Have the Tools It Needs?, Washington, DC: Project on Emerging Technologies, Woodrow Wilson International Centre for Scholars. Available at: <http://eprints.internano.org/62/1/Food_Packaging_pen12.pdf>.

Taylor, M. R., 2008. Assuring the safety of nanomaterials in food packaging: the regulatory process and key issues, Technical Report, Woodrow Wilson International Center for Scholars.

Teubner, G., 1987. Juridification: concepts, aspects, limits, solutions. In: Teubner (Ed.), Juridification of Social Spheres: A Comparative Analysis in the Areas of Labour, Corporate, Antitrust and Social Welfare Law. W. de Gruyter, Berlin, p. 15.

The Royal Society and The Royal Academy of Engineering, July 2004. Nanoscience and nanotechnologies: opportunities and uncertainties. Available at <http://royalsociety.org/Nanoscience-and-nanotechnologies-opportunities-and-uncertainties-/>; <http://www.euractiv.com/en/science/nanoscience-nanotechnologies-opportunities-uncertainties/article-133080>; <http://www.bis.gov.uk/files/file34431.pdf>.

van der Meulen, B.M.J., 2007. Regulating GM food: three levels, three issues. In: Somsen, H. (Ed.), The Regulatory Challenge of Biotechnology. Human Genetics, Food and Patents. Edward Elgar, Cheltenham, UK, pp. 139–156.

van der Meulen, B.M.J., 2009a. Science based food law. Eur. Food Feed Law 1, 58–71.

van der Meulen, B.M.J., 2009b. The system of food law in the European Union, Deakin Law Rev., 14: 305–339. Available at: <http://www.deakin.edu.au/buslaw/law/dlr/pdf_files/vol14-iss2/7vandermeulen.pdf>.

van der Meulen, B.M.J., 2009c. Reconciling Food Law to Competitiveness. Wageningen Academic Publishers, Wageningen, The Netherlands.

van der Meulen, B.M.J., 2012. The structure of European Food Law. Laws 2, 69–98. 10.3390/laws2020069.

van der Meulen, B.M.J., Bremmers, H.J., Wijnands, J.H.M., Poppe, K.J., 2012. Structural precaution: the application of premarket approval schemes in EU food legislation. Food. Drug. Law. J. 67, 453–473.

Vogel, D, 1995. Trading Up: Consumers and Environmental Regulation in a Global Economy. Harvard University Press, USA.

Whatmore, R.W., 1999. Ferroelectrics, microsystems and nanotechnology. Ferroelectrics 225 (1), 179–192. Available at: <http://www.informaworld.com/smpp/content~content = a752139656&db = all>.

Zijverden, M., Sips, A.J.A.M., 2008. Nanotechnology in perspective. RIVM-report nr. 601785002, <http://www.rivm.nl/bibliotheek/rapporten/601785003.pdf>.

Zweigert, K., Kötz, H., 1996. Einführung in die Rechtsvergleichung, third ed. Mohr Siebeck, p. 45.

MISCEALLEANOUS

Center for Responsible Nanotechnology, <http://www.crnano.org/whatis.htm>.

Safe Foods, <http://www.safefoods.nl/default.aspx>.

GLOSSARY

For the purpose of this study the following expressions are given the following meaning

Food   Any substance, whether processed, semi-processed or raw, which is intended for human consumption, and includes ingredients, additives, drink, chewing gum and any substance which has been used in the manufacture, preparation or treatment of "food" but does not include cosmetics or tobacco or substances used only as drugs[1]

Food law   Regulatory framework for the food sector

Nanofoods   Products intended or reasonably expected to be ingested by humans that are likely to contain substances obtained by nanoscale chemosynthesis or engineered nanoparticles

Regulatory framework   The entire hierarchy of laws, regulations, guidelines, codes of conduct, implementing policies, scientific policies, conformity assessment requirements, powers of inspection, sanitation and sanctioning and any other tools (by whatever name) that are in place to deal with certain issues such as food safety

Sponsor   A business responsible for placing a nanofood on the market, such as a producer, exporter, importer, trader, brand owner or private label retailer

---

[1]This definition has been adopted from the Codex Alimentarius. See Codex Alimentarius Commission. Procedural Manual, 21st edition, Rome 2013, p. 22.

www.ingramcontent.com/pod-product-compliance
Lightning Source LLC
Chambersburg PA
CBHW060444240326
41598CB00087B/3481